Collector's Guide to the
VESUVIANITE GROUP

Schiffer Earth Science Monographs Volume 5

Robert J. Lauf

4880 Lower Valley Road, Atglen, Pennsylvania 19310

Other Schiffer Books by Robert J. Lauf
Collector's Guide to the Axinite Group.
ISBN: 9780764332166. $19.99.
Collector's Guide to the Epidote Group.
ISBN: 9780764330483. $19.99.
Collector's Guide to the Mica Group.
ISBN: 9780764330476. $19.99.
Introduction to Radioactive Minerals.
ISBN: 9780764329128. $29.95.

Other Schiffer Books on Related Subjects
Collecting Fluorescent Minerals. Stuart Schneider.
ISBN: 0764320912. $29.95.
Collector's Guide to Fluorite. Arvid Eric Pasto.
ISBN: 978-0-7643-3193-0. $19.99.
Gems & Minerals. Dr. Andreas Landmann.
ISBN: 9780764330667. $29.99.
The World of Fluorescent Minerals. Stuart Schneider.
ISBN: 0764325442. $29.95.

Copyright © 2009 by Robert J. Lauf
Library of Congress Control Number: 2008941377

All rights reserved. No part of this work may be reproduced or used in any form or by any means—graphic, electronic, or mechanical, including photocopying or information storage and retrieval systems—without written permission from the publisher.

The scanning, uploading and distribution of this book or any part thereof via the Internet or via any other means without the permission of the publisher is illegal and punishable by law. Please purchase only authorized editions and do not participate in or encourage the electronic piracy of copyrighted materials.

"Schiffer," "Schiffer Publishing Ltd. & Design," and the "Design of pen and ink well" are registered trademarks of Schiffer Publishing Ltd.

Designed by Mark David Bowyer
Type set in Arno Pro / Humanist521 BT

ISBN: 978-0-7643-3215-9
Printed in China

Schiffer Books are available at special discounts for bulk purchases for sales promotions or premiums. Special editions, including personalized covers, corporate imprints, and excerpts can be created in large quantities for special needs. For more information contact the publisher:

Published by Schiffer Publishing Ltd.
4880 Lower Valley Road
Atglen, PA 19310
Phone: (610) 593-1777; Fax: (610) 593-2002
E-mail: Info@schifferbooks.com

For the largest selection of fine reference books on this and related subjects, please visit our web site at
www.schifferbooks.com
We are always looking for people to write books on new and related subjects. If you have an idea for a book please contact us at the above address.

This book may be purchased from the publisher.
Include $5.00 for shipping.
Please try your bookstore first.
You may write for a free catalog.

In Europe, Schiffer books are distributed by
Bushwood Books
6 Marksbury Ave.
Kew Gardens
Surrey TW9 4JF England
Phone: 44 (0) 20 8392-8585; Fax: 44 (0) 20 8392-9876
E-mail: info@bushwoodbooks.co.uk
Website: www.bushwoodbooks.co.uk
Free postage in the U.K., Europe; air mail at cost.

Contents

Preface 5

Acknowledgments 7

Introduction 8
    Table 1: Obsolete and Varietal Names for Vesuvianite 9

Taxonomy of the Vesuvianite Group 13
    General Formula 13
    Table 2: Minerals of the Vesuvianite Group 14
    Structure and Habit 14
    Chemistry and Color 25

Formation and Geochemistry 32
    Vesuvianites in Igneous Rocks 32
    Vesuvianites in Metamorphic Rocks 32

The Minerals 42
    Vesuvianite 42
    Fluorvesuvianite 79
    Manganvesuvianite 80
    Wiluite 84

References 89

Preface

This volume continues a series of monographs on important groups of so-called rock forming silicates, the purpose of which is to help mineral collectors gain a better appreciation of these complex minerals. Because of the importance of rock forming minerals in geological processes, they are the subject of extensive published research, much of which has been brought together in the five-volume compendium *Rock-Forming Minerals* (Deer, Howie, and Zussman 1962) and the greatly expanded Second Edition thereof. Among rock-forming minerals, the vesuvianite group is perhaps best known to collectors through the colorful crystals that have been collected at the Jeffrey quarry, Asbestos, Canada. Spectacular finds in China and Pakistan have added to the interest in this mineral group. The "new" species wiluite was known informally for several hundred years (and considered to be a variety of vesuvianite) before recently being elevated to species status. Other recently-described species are manganvesuvianite and fluorvesuvianite. Vesuvianite is occasionally cut as a gem, making it of some interest to lapidaries. The present monograph, which grew out of a paper by the author for *Rocks & Minerals* (Lauf 2009), is organized as follows: After a brief introduction, the general treatment begins with an explanation of the chemistry and taxonomy of the group and a discussion of ongoing research into problems such as optically anomalous crystals. A section on their formation and geochemistry explains the kinds of environments where vesuvianites are formed. Then, a detailed entry for each mineral provides information on important localities and full-color photos wherever possible so that collectors can see what good specimens look like and which minerals one might expect to find in association with vesuvianites. As in earlier volumes in this series, the photographs were not selected to showcase extremely expensive "museum pieces" or purported "best in the world" specimens, but instead, good specimens that an interested collector could actually hope to obtain and study.

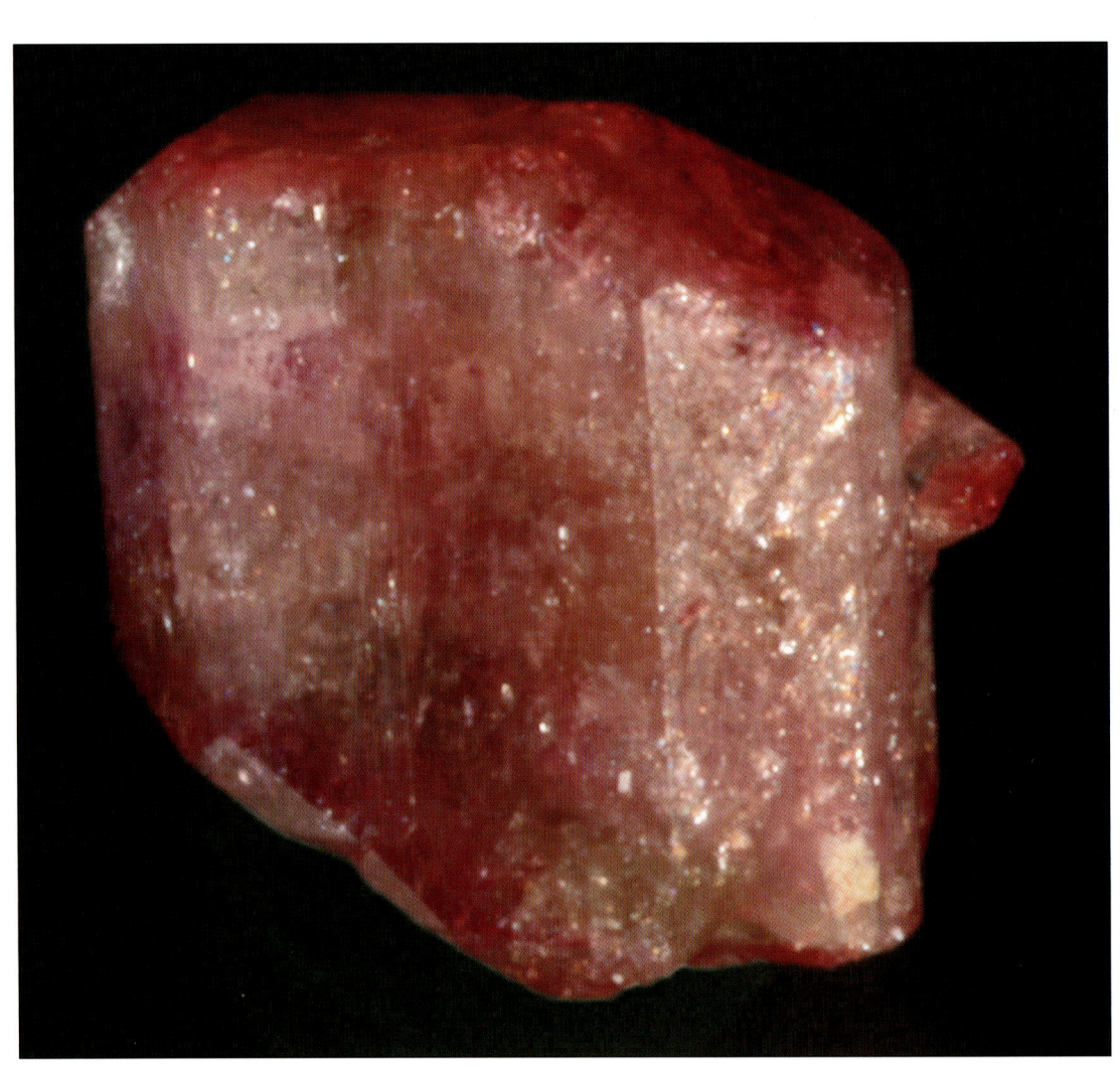

Acknowledgments

The following colleagues kindly provided technical information, literature, and helpful discussions: Thomas Armbruster, *University of Bern*, Switzerland; Deborah Cole, *Oak Ridge National Laboratory*; Evgeny Galuskin, *University of Silesia*, Poland; Lee Groat, *University of British Columbia*, Canada; Peter Leavens, *University of Delaware*. Important specimens and background information were supplied by Dudley Blauwet, *Mountain Minerals*; Dave Bunk; Sharon Cisneros, *Mineralogical Research Co.*; Richard Dale, *Dale Minerals*; Kevin Downey, *Well-Arranged Molecules*; Jordi Fabre, *www.fabreminerals.com*; Shields Flynn, *Trafford-Flynn Minerals*; Leonard Himes and Michael Jacob, *Minerals America*; Danny Jones, *Firebird Minerals*; Patrick Kelley, *PAK Designs*; Rich Kelly, *South Fork Mining*; and Tony Nikischer, *Excalibur Mineral Co.*

Introduction

The vesuvianite group presently comprises four minerals: vesuvianite, fluorvesuvianite, manganvesuvianite, and wiluite. Good crystals are found throughout the world, in a variety of forms and colors, and vesuvianite is also a minor gem mineral. Not only does the group include many superb specimens for the collector, but the four species and their varieties are also the subject of intense research interest because of their complicated chemistry and structure (Groat, Hawthorne, and Ercit 1992a; Groat et al. 1993). The diversity of colors and habits has led several authors to refer to vesuvianite as a "mineral chameleon" (Arem 1973; Gnos and Armbruster 2007).

Vesuvianite nomenclature has a long and troubled history. A generation ago Arem (1973) tabulated twenty names that had been applied to the mineral since 1723, along with several varietal names; Arem recognized that vesuvianite had priority, and yet in that paper he continued to refer to the mineral as "idocrase" based on popular use and indecision by the International Mineralogical Association (IMA). Obsolete and varietal names are given in Table 1.

According to Arem (1977), "*Idocrase* is one of the lesser known and more beautiful collector gems. When properly cut it is as bright and attractive as the garnets which it so strongly resembles (grossular)." Faceting-grade material is known from a handful of localities, but the overall size of the crystals (or the amount of transparent material in the larger crystals) generally limits good stones to less than about 10-15 carats.

An interesting historical account is given by Bauer (1904): "Because of its occurrence on Mount Vesuvius idocrase is frequently referred to as vesuvian or vesuvianite. It occurs at this locality in remarkably fine transparent brown crystals, which are sometimes cut at Naples as gems, and, on this account, are known in the trade as 'Vesuvian gems.' The use of this stone as a gem is not extensive, and is mainly confined to Italy. Crystals of gem-quality and of a green color are found also in the Ala valley in the Piedmontese Alps; a small number find their way into the gem markets through the neighboring town of Turin. Idocrase from other sources is scarcely ever cut as a gem, so that it may be regarded as an Italian precious stone."

Faceting-grade green vesuvianite has been found at several places in East Africa, particularly the Kallad/Namanga area of Kenya (Keller 1992). Cut stones from this locale are occasionally seen in the gem trade.

Figure 1. Vesuvianite gemstones: a rectangular golden brown stone from Canada and an oval green stone from Kenya.

Table 1. Obsolete and Varietal Names for Vesuvianite

Name	Definition
Californite	A compact green mixture of vesuvianite and grossular
Cyprine	Blue vesuvianite containing copper
Duparcite	Vesuvianite from Morocco
Egeran	Brown or greenish vesuvianite from Eger River area, Czech Republic
Frugardite	Vesuvianite
Genevite	Vesuvianite from Morocco
Gokumite	Vesuvianite from Gokum, Sweden
Heteromerite	Vesuvianite
Idocrase	Vesuvianite
Jewreinowite	Vesuvianite
Loboite	Vesuvianite from Gokum, Sweden
Manganidocrase[a]	Vesuvianite containing some Mn from Jordansmuhl, Poland
Titanvesuvianite	Vesuvianite containing some titanium
Xanthite	Yellow vesuvianite from Amity, New York

[a]This material contains a small amount of Mn and *is not* the same as the accepted species manganvesuvianite.

Massive green vesuvianite, usually containing some grossular or other mineral, is found in several places in California and is often called *Californite* or *California jade*. *Californite* has long been used as a jade substitute, and the best gem-grade material has excellent rich green color and takes a high polish. An entertaining history of the re-discovery of the deposit at Happy Camp, California, is given by Kraft (1947), the noted industrialist who eventually bought the claim. At present, several claim holders operate in the area, working both in-situ and alluvial deposits.

Figure 2. A hand-sized gem-grade *californite* rough found in the Klamath River at the South Fork Mining claim (top), and a polished freeform shape (bottom) showing the rich green color and excellent luster seen in the best material.

Figure 3. A chunk of "specimen grade" *californite* from the South Fork Mining claim (left), and a tumble-polished piece of the same material (right). Here, a softer phase, possibly serpentine, prevents the material from taking a high polish. Comparing this material to that in the previous photo, one can see that lapidaries who have only encountered lower-grade material might have a poor opinion of *californite*.

12   Introduction

Figure 4. A simple pendant in sterling silver features a 15 × 25 mm cabochon of *californite*.

Figure 5. A *californite* carving, 8 cm long and 4 cm tall, fabricated in Thailand of material from the South Fork Mining claim.

Taxonomy of the Vesuvianite Group

General Formula

Members of this group are tetragonal silicates (space group P4/nnc) whose general formula may be summarized as follows:

$X_{19}Y_{13}Z_{18}T_{0-5}O_{68}W_{10}$ where:
X is Ca, Na, REE^{3+}, Pb^{2+}, Sb^{3+};
Y is Al, Mg, Fe^{3+}, Fe^{2+}, Ti^{4+}, Mn, Cu, Zn;
Z is Si;
T is B; and,
W is (OH, F, O).

The chemical formulas for the four accepted species are given in Table 2.

The general formula presented here is that proposed by Groat et al. (1992a), who analyzed seventy-six vesuvianites, representing about fifty localities, and evaluated many existing formulas with the goal of developing a formula that best fits the known range of compositions. In this scheme, the X cations occupy [8]-coordinated sites; the Y cations occupy [6]- and [5]-coordinated sites; the Z cations occupy [4]-coordinated (tetrahedral) sites; and W are monovalent or divalent anions. Cations on the X sites are predominantly divalent (mostly Ca), and those occupying the Z sites are tetravalent (and almost exclusively Si). Polyvalent substitutions are largely confined to the Y sites, where di-, tri-, and even tetravalent ions can reside.

The distribution of cations over the possible sites (X, Y, and Z) may be surmised from a large group of microanalyses by assigning cations to the different sites and checking to see if the cation sums for each group of sites show normal distributions. At the same time, the assignments need to make sense in terms of general ionic size and charge considerations. Based on this type of analysis, several general observations can be made:

The Z sites are all tetrahedrally coordinated and are dominantly or exclusively occupied by Si. Specifically, there is little or no substitution of Al on these sites.

The X sites are [8]- or [9]-coordinate and are dominantly occupied by Ca, as indicated by the nominal formulas for the individual minerals. Other cations that are assigned to these sites when present include Na, REE, Pb, Bi, and Th.

The Y sites are dominantly occupied by Al, along with significant Mg and Fe and minor Ti. Other transition metals (e.g., Cr^{3+}, Mn, Zn, Cu^{2+}) likely occupy Y sites as well.

In vesuvianites containing significant amounts of boron, it was not clear where the B would reside, but substitution of (BO$_4$) for (SiO$_4$) was considered unlikely. Detailed crystallographic analysis of samples with up to about 4% B$_2$O$_3$ identified two new sites (designated T in the general formula) and suggested that boron is accommodated primarily by the substitution B^{3+} + Mg^{2+} ↔ 2H$^+$ + Al^{3+} (Groat et al. 1994).

In some vesuvianites, the number of cations that are normally assigned to the Y sites (Al, Mg, Fe^{2+}, Fe^{3+}, Ti^{4+}, Cu^{2+}, Zn, and Mn) significantly exceeds thirteen, the number of available sites in the structural formula. Crystallographic analysis of some of these specimens indicated that excess Al or Fe occupies the T(1) site, compensated by a vacancy on an X site (Groat et al. 1994a).

Table 2. Minerals of the Vesuvianite Group

Species	Formula
Vesuvianite	$Ca_{19}(Al,Mg,Fe^{2+})_{13}Si_{18}O_{69}(OH,F)_9$
Fluorvesuvianite	$Ca_{19}(Al,Mg,Fe^{2+})_{13}(SiO_4)_{10}(Si_2O_7)_4O(F,OH)_9$
Manganvesuvianite	$Ca_{19}Mn^{3+}(Al,Mn^{3+},Fe^{3+})_{10}(Mg,Mn^{2+})_2Si_{18}O_{69}(OH)_9$
Wiluite	$Ca_{19}(Al,Mg,Fe,Ti)_{13}(B,Al,\square)_5Si_{18}O_{68}(O,OH)_{10}$

Structure and Habit

Early structural models for vesuvianite recognized the close similarities between vesuvianite and grossular, and explained the structure in terms of "garnet-like" columns parallel to the c-axis. The structure was assigned to space group P4/nnc and this generally fit the observed X-ray diffraction data for samples from many localities. However, it has been known for quite some time that XRD patterns for some vesuvianites show violations of the criteria for P4/nnc, suggesting slightly different structures (Arem and Burnham 1969).

More recently, the basic crystallography of the vesuvianite group has been studied extensively (Rucklidge et al. 1975; Groat et al. 1992; Groat et al. 1993; Groat et al. 1996; Fitzgerald, Rheingold, and Leavens 1986; Yoshiasa and Matsumoto 1986; Ohkawa, Yoshiasa, and Takeno 1992; Allen and Burnham 1992; Armbruster and Gnos 2000; Groat, Hawthorne, and Ercit 1992b; Groat, Hawthorne, and Ercit 1994; Groat, Hawthorne, and Ercit 1994a). As noted, the vesuvianite structure ideally has P4/nnc symmetry, but the physical properties seen in many samples indicate deviations from this. Based on their optical properties, samples could be placed into three groups: (1) normal crystals that are optically uniaxial, as expected for tetragonal symmetry; (2) blocky crystals with irregularly shaped areas that are optically biaxial with variable birefringence and 2V values in the range of 5-35°; and (3) optically sector-zoned biaxial crystals showing {001}, {101}, and {100} sectors with different optical orientation and low (about 5°), intermediate (20-35°), and high (40-60°) values of 2V respectively. Combining optical and X-ray data indicated that the symmetry of biaxial crystals might be P2/n or Pn. It was suggested that there is a continuous or nearly continuous ferroelastic phase transition between high-temperature P4/nnc structure and a lower temperature P2/n or Pn structure. The various optical types could be the result of different relationships between the temperature of crystallization and the temperature of the phase transition (Groat et al. 1993).

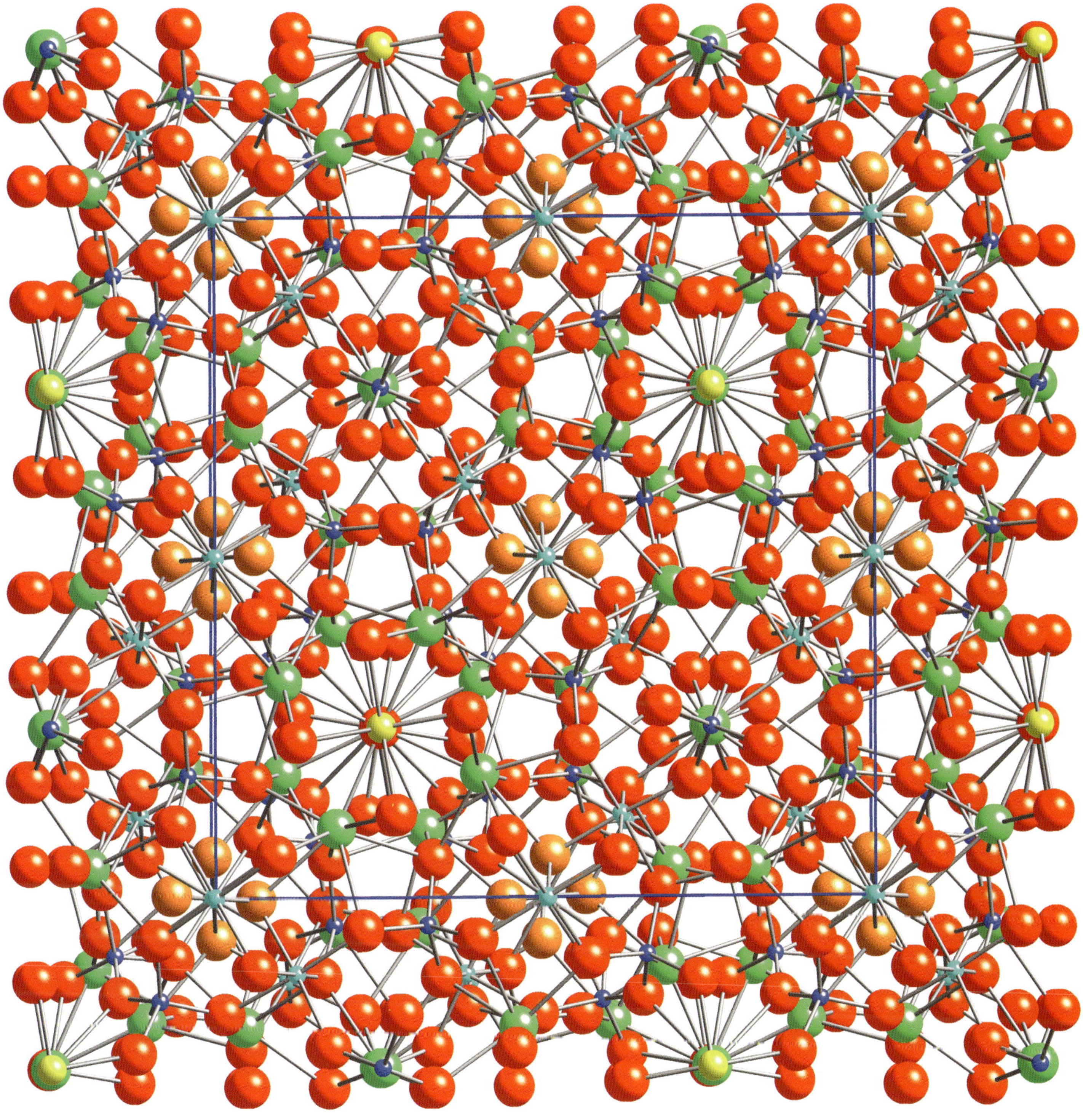

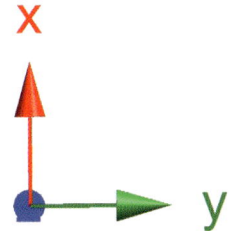

Figure 6. The crystal structure of vesuvianite. In this "ball and stick" model, red balls are oxygen, green are Ca, dark blue are Si, yellow are Mg, and gold are OH⁻. The structure is viewed along the Z axis, and the unit cell is outlined in blue. Some of the surrounding atoms are also shown in order to better visualize the periodicity of the lattice.

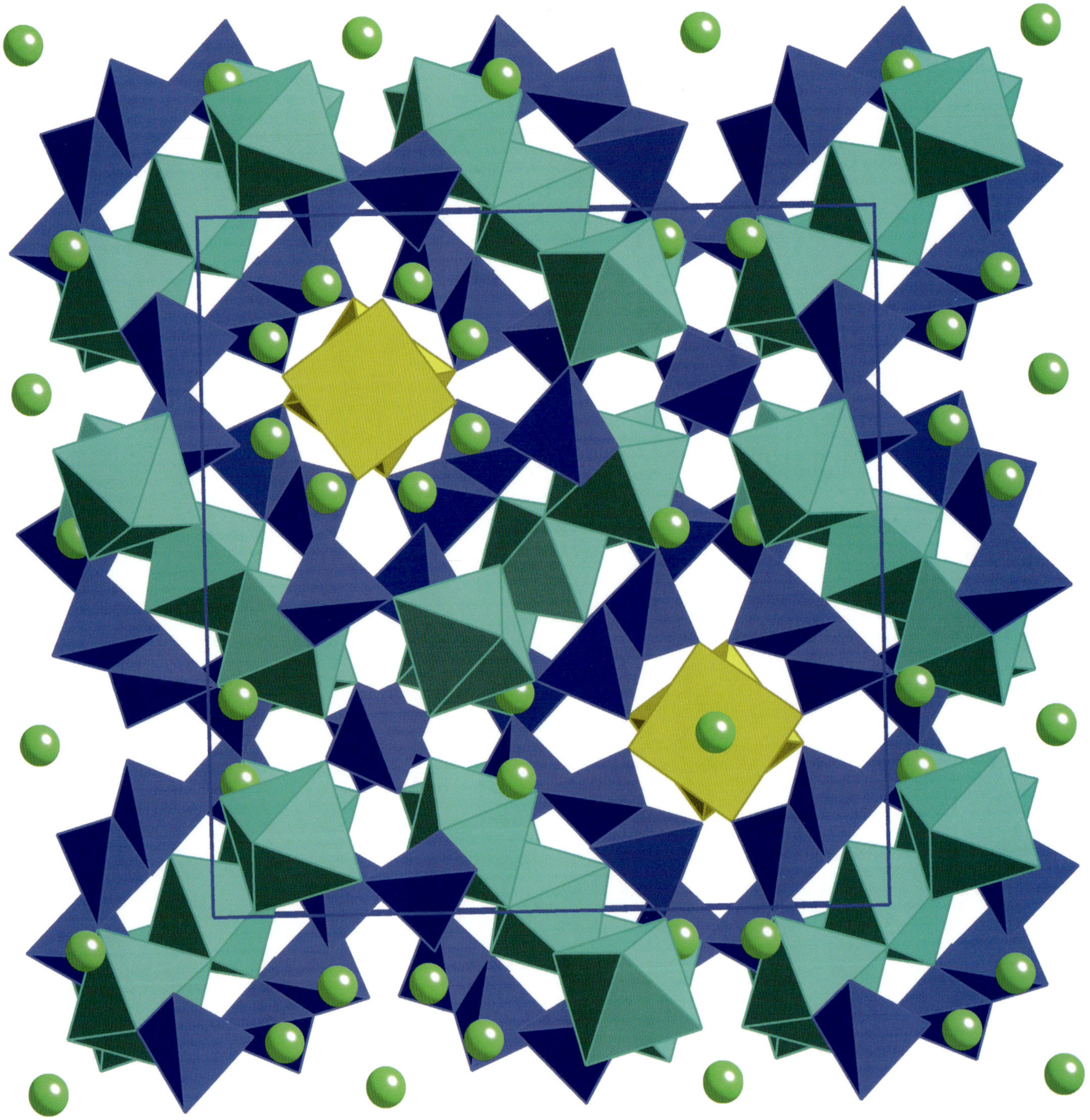

16 *Taxonomy of the Vesuvianite Group: Structure and Habit*

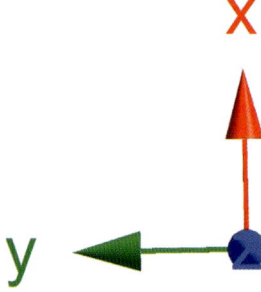

Figure 7. The same structure and view as the previous figure, but shown as a polyhedral model. Here, a polyhedron denotes a metal ion surrounded by the oxygens to which it is coordinated; the metal is hidden within the polyhedron and the oxygen positions are represented by its corners. Thus, for example, the dark blue tetrahedra represent SiO_4 groups and so forth. For better clarity, Ca atoms are shown as balls rather than as coordination polyhedra.

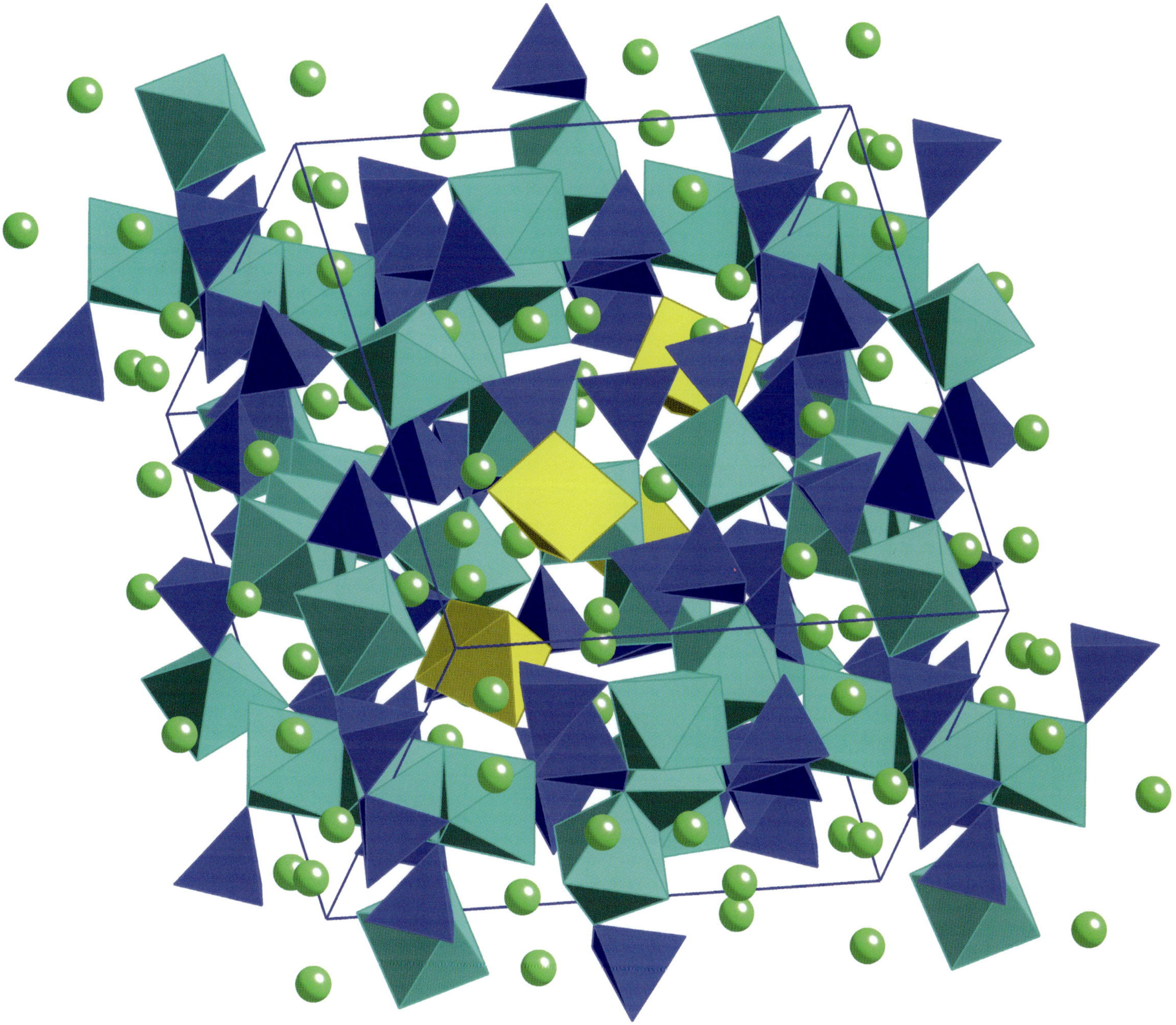

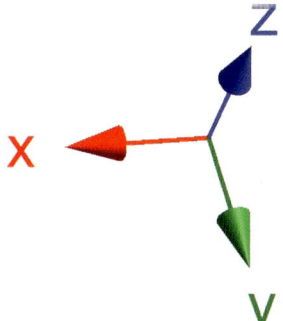

Figure 8. The structure shown in the previous figure but rotated to view the unit cell diagonally, in order to better visualize the three-dimensional structure of the crystal.

Sectored vesuvianite from Bellecombe, Italy, that showed anomalous triclinic optical properties was studied by X-ray diffraction and polarized Fourier transform IR spectroscopy (Tanaka, Akizuki, and Kudoh 2002). The results suggested that the optical effects were not explained by phase transitions or residual stresses, but rather seemed to be produced during non-equilibrium crystal growth.

Vesuivianite grains in a calc-silicate rock from the Canigou Massif, France, show oscillatory concentric birefringence-zoning, with up to nine dark (low-birefringence) zones alternating with lighter-colored zones. Some zones are as thin as 10 μm wide. Electron microprobe traces across these crystals were used to correlate compositional fluctuations with the observed optical zoning, leading to the following observations (Gibson, Wallace, and de Bruin 1995):

a. The vesuvianite grains are zoned with respect to Ti, Al, Mg, and Fe.

b. There is a positive correlation between the intensity of birefringence and the concentration of Ti.

c. Cation zoning also occurs within optically homogeneous birefringent zones and appears to reflect more than one substitution.

d. Substitutions operative in low-birefringence zones appear to be Ti + Mg ↔ 2Al and Mg ↔ Fe^{2+}, whereas in high-birefringence zones the substitutions appear to be Fe^{3+} ↔ Al and Ti + Mg ↔ 2Al.

To further complicate things, some low-temperature vesuvianites exhibit long-range ordering leading to P4/n or P4nc symmetry and a phenomenon known as "rod polytypism." Briefly, within the structure there are "strings" along the four-fold axes that can be arranged in several ways. Each string is fully ordered (in the short range) but adjacent strings can either be long-range disordered leading to a P4/nnc space group, or exhibit some particular ordering patterns that lead to lower overall symmetry (Armbruster and Gnos 2000). By studying a number of specimens from metamorphic rocks that had formed under fairly well understood conditions, it was possible to relate the symmetry and rod arrangement to the crystallization temperature. Specifically, P4/nc-dominant structures are characteristic of temperatures below 300°C; P4/n-dominant structures about 300 - 500°C; and P4/nnc-dominant structures above 500°C (Gnos and Armbruster 2006; Gnos and Armbruster 2007).

High-resolution transmission electron microscopy on vesuvianite from the skarn at Crestmore quarry, Crestmore, California, produced electron diffraction spots indicating violations of space group P4/nnc and consistent with P4/n. Using dark-field as well as high-resolution images revealed a domain structure in which the domains were generally 10 - 50 nm wide and elongated parallel to [001]. The structure was interpreted to have formed by transformation twinning that resulted from ordering of the disordered high-temperature P4/nnc structure (Veblen and Wiechmann 1991).

These fine structural details are not necessarily of direct interest to the average mineral collector, but they are central to the study of petrology, because accurately inferring *how* a particular deposit formed and *at what temperature* is crucial to understanding the local geology [see, for example, Gnos and Armbruster (2006)].

Although optically anomalous vesuvianites, i.e., those that are biaxial rather than uniaxial, are of lower symmetry (and thus have subtly different structures) than tetragonal crystals, presumably attributed to cation ordering, the IMA has not yet issued guidelines for treating these materials as potentially new species.

Vesuvianites display a fascinating variety of crystal morphologies and habits. Goldschmidt (1916) presented 249 drawings of natural vesuvianite crystals. A few examples, Figure 9, have been selected to show some commonly seen types. The fairly simple tetragonal prism, upper left, is characteristic of vesuvianite from Lake Jaco, Mexico, and of wiluite from Russia. The equant crystal terminated with a large, flat (001) face, upper right, illustrating a sample from the type locale at Mount Vesuvius, typifies a habit seen at Sanford, Maine, and several localities in Pakistan and Italy. An elongated prism with a more acute termination bounded by {131} faces, center left, can be found in Alpine cleft-type localities in Europe and Pakistan, as well as at Mount Belvidere, Vermont. The simple elongated prismatic crystal, center right, illustrating a sample from Drammen, Norway, is also fairly typical of manganvesuvianite crystals. Equant tetragonal prisms with intricate surface figures, bottom left and right, have been noted in some wiluites from Russia.

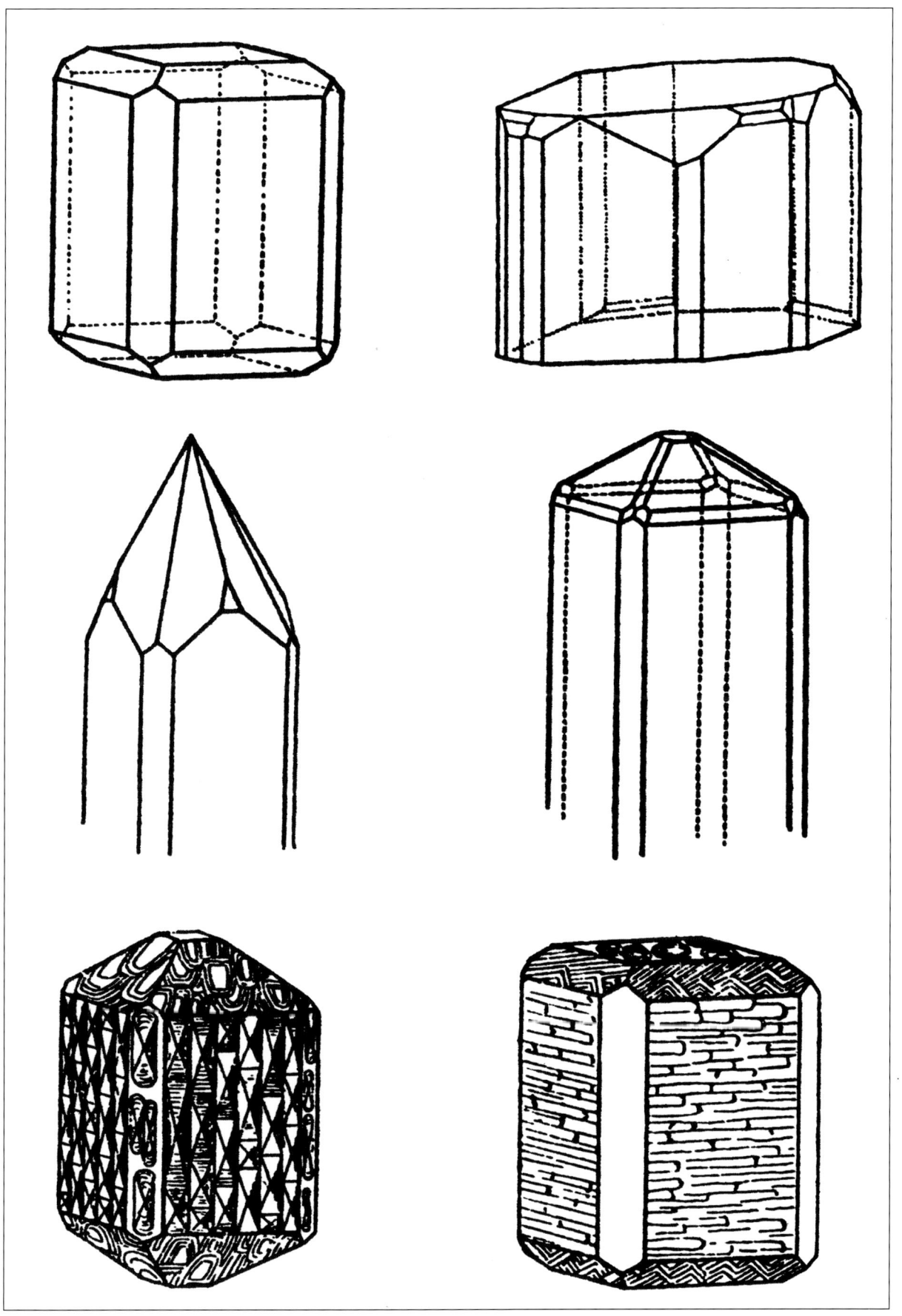

Figure 9. Drawings of some natural vesuvianite crystals, modified from Goldschmidt (1916).

Figure 10. An intergrown pair of simple prismatic crystals from Lake Jaco, Coahuila, Mexico. *RJL8*

Figure 11. Equant crystal with vertical striations and a prominent (001) termination, from California. *RJL3344*

Figure 12. Two dull green "floater" crystals from Pakistan; it is interesting that although each crystal is doubly-terminated, the terminations on the two ends are not identical. *RJL3341 and RJL3342.*

Figure 13. Elongated prismatic crystals of manganvesuvianite from the N'Chwaning II mine, South Africa. *RJL2768*

A rare but important morphological type, not known to Goldschmidt, is the whisker habit. Fluorvesuvianite has in fact only been found as whiskers. Studies of high-fluorine vesuvianite whiskers in skarns from Polar Yakutia, Russia, concluded that whiskers form under kinetic conditions in which the growth of prism faces {100} and {110} is inhibited by poisoning of surface sites by surface active species (e.g., manganese), so that the crystal grows much faster along (001) thereby creating a rod or needle (Galuskin et al. 2003a).

Figure 14. Fluorvesuvianite showing the whisker habit, which is somewhat anomalous for vesuvianites. *RJL2830*

Chemistry and Color

Depending on the amount and type of chromophores present in a crystal, vesuvianite can be colorless to pale yellow, pale pink to deep red-purple (Mn^{2+}, Mn^{3+}), yellowish-green (Fe^{2+}) to rich emerald-green (Cr^{3+}), bright blue (Cu^{2+}) to blue-green (Cu^{2+} and Fe^{2+}), and brown to black (Ti^{4+}). Chromophores can often impart significant color even at fairly low concentrations. For example, the green "chrome" vesuvianite from the iron-bearing serpentinite at Vermion, Greece, contains about 0.7% Cr_2O_3 and brilliant blue *cyprine* from Telemark, Norway, contains about 0.8% CuO. Small fluctuations in chemistry as the crystal grows can occasionally produce very dramatic color zoning, seen perhaps most clearly in the pink-yellow-green crystals from the Jeffrey quarry, Quebec, Canada.

Ti-rich vesuvianites were studied using polarized absorption spectroscopy, which showed that their color and pleochroism arise from $Fe^{3+} \rightarrow Ti^{4+}$ and $O^{2-} \rightarrow Fe^{3+}$ charge transfer processes. In green vesuvianite with low Ti, over 90% of the iron is present as Fe^{3+} on octahedral (**Y**) sites (Manning 1975; Manning and Tricker 1975).

Manning (1977) proposed the following loose classification of vesuvianites in terms of transition-metal chemistry and color:

a. Blue, pink, and lilac crystals in which color originates primarily in *d-d* transitions in Cu^{2+}, Mn^{3+}, and Cr^{3+}.

b. Green crystals in which the main chromophore is apparently Fe^{3+} on Al/Fe sites. Examples include crystals from Lowell, Vermont, from Sanford, Maine, and from Pakistan. The $O^{2-} \rightarrow Fe^{3+}$ charge transfer process is suppressed, perhaps by a high OH content; it is noted that the material from Lowell, Vermont contains about 2% water.

c. Fe^{2+}-rich crystals in various shades of brown and yellow. The crystals can be Ti-rich (> 1% TiO_2) or Ti-poor (< 0.2% TiO_2). Fe^{2+} occupies mainly **Y** sites, with fairly little substituting for Ca on **X** sites.

d. Crystals that may contain transition-metal ions in sites of unusual coordination, such as Fe^{3+} in 5- or 8-coordination, and in large amounts up to 10 – 20% of total metals; these substitutions may be coupled to others. These materials, containing both Fe^{3+} and Fe^{2+} ions, at least partly occupying 5-coordinated **Y** sites, are usually brown.

Figure 15. A colorful zoned vesuvianite crystal from Jeffrey quarry. *RJL3277*

In another study of chemical variation, 44 vesuvianites of various colors, morphologies, and occurrences were analyzed (Fitzgerald, Leavens, and Nelen 1992) and placed into four types somewhat analogous to the categories proposed by Manning (1977); it was concluded that "Color provides a good, approximate basis for categorizing vesuvianites in the absence of chemical information."

Figure 16. A group of thumbnail-sized vesuvianites from Jeffrey quarry, Asbestos, Quebec, showing the range of colors and habits found there.

Figure 17. Intense purple Mn-containing vesuvianite from Jeffrey quarry, Asbestos, Quebec. Sample is just over 3 cm across. *RJL2665*

Figure 18. Rich green vesuvianite from Jeffrey quarry, Asbestos, Quebec, showing the influence of chromium on the color of the crystals. *RJL3262*

Figure 19. Bright blue vesuvianite var. *cyprine* from Telemark, Norway. *RJL3282*

Figure 20. Green chromian vesuvianite from the Yerington district, Nevada. *RJL2712*

Figure 21. An unusual black vesuvianite from the Bill Waley Allotment tungsten mine, Tulare County, California. *RJL3381*

Formation and Geochemistry

Minerals of the vesuvianite group principally occur in contact metamorphic zones in limestones (e.g., skarns) where they are associated with garnet, diopside, and wollastonite; they are also found in regionally metamorphosed limestones. Vesuvianite has also been noted in veins associated with mafic igneous rocks and serpentinites, in garnetized gabbros (known as rodingites), and in nepheline-syenites (Deer, Howie, and Zussman 1982).

Vesuvianites in Igneous Rocks

Vesuvianite occurrences in alkaline igneous rocks such as nepheline syenite include Almunge, Sweden; Iron Hill, Colorado; Seiland, Norway; and the Fukushinzan district, Korea (Deer, Howie, and Zussman 1982). An unusual metamict U- and Th-bearing vesuvianite was described from a syenite in the Kachauik pluton, Seward Peninsula, Alaska (Himmelberg and Miller 1980). Metamict vesuvianite from the Tuva district, Russia, occurs in alkaline pegmatite veins associated with nepheline, microcline, thomsonite, cancrinite, and albite (Kononova 1960). Vesuvianite is "fairly common" at Mont Saint-Hilaire (Mandarino and Anderson 1989), where it is found in xenoliths within the nepheline syenite but more commonly in the contact zone between marble and the syenite; thus most of the vesuvianite at Mont Saint-Hilaire is more properly regarded as metamorphic rather than as a primary component of the igneous rock itself.

Vesuvianites in Metamorphic Rocks

Vesuvianite is stable under both reducing and oxidizing conditions because variable amounts of di- and trivalent cations can be accommodated in its structure. It is stable at temperatures from less than 300°C to more than 900°C, depending on pressure. In metamorphic rocks vesuvianite tends to be associated with different mineral assemblages depending on metamorphic grade (Gnos and Armbruster 2006):

In low grade metamorphic environments (<300°C), major associated minerals include hydrogarnet, xonotlite, wollastonite, natrolite, thomsonite, pectolite, and calcite. Other species that might be present include kirschsteinite, glaucochroite, aegirine, åkermanite, tobermorite, hydromagnesite, and vuagnatite.

At medium grades (~300–500°C) the associated minerals include grossular (particularly var. *hessonite*), diopside, Mg-chlorite, prehnite, epidote, and calcite. Other species might include albite, titanite, apatite, dolomite, perovskite, and zircon.

At high grades (>500°C) the associated minerals include grossular, diopside, wollastonite, calcite, monticellite, melilite, and quartz. Other species might include spinel, epidote, plagioclase, and zoisite.

Skarns and other contact metamorphic rocks represent many classic vesuvianite locales. At the type locality at Mt. Vesuvius, the mineral is found in limestone blocks ejected from the volcano. The "xenoliths" have undergone intense thermal metamorphism (600 – 1050°C, 0 - 2 kbar). A similar process occurred at Ariccia, Italy, where wiluite crystals are found in skarn-type volcanic ejecta, which are considered to be fragments of metasomatized carbonate wall rock (Bellatreccia et al. 2005). The skarns near Lake Jaco, Coahuila, Mexico, are the result of fairly high-grade metamorphic conditions (> 600°C, 1 – 3 kbar).

Figure 22. Equant brown vesuvianite crystals about 1 cm associated with reddish-brown grossular, illustrating the products of intense metamorphic alteration in skarn at the type locale, Mt. Vesuvius, Italy. *RJL3374*

Figure 23. Striated prismatic vesuvianite crystals typical of a skarn deposit, from Sanford, Maine. Note the white massive calcite in the area between the crystals. *RJL3270*

Figure 24. A large, lustrous brown crystal with a smaller crystal fragment in skarn from Morocco. *RJL3389*

Excellent green and brown crystals to 10 cm are found in Grenville Province skarns in Ontario, Canada, particularly at the Pinchon marble quarry near Malone (Robinson and Chamberlain 1982). In the Wilui River region, Russia, wiluite is formed in the contact aureole of Siberian Trapp dikes (> 600°C, <1 kbar). It has been suggested that the mineral assemblage represents a skarn that was later serpentinized, altering all of the existing skarn minerals except wiluite and grossular (Groat et al. 1998). Skarns containing up to 2 percent vesuvianite are found in the Italian Mountain area, near Aspen, Colorado (Treube 1984). In tin-bearing skarns in eastern Siberia, vesuvianite in various shades of green and brown contains small amounts of tin (up to about 90 ppm). The brown samples are distinguished from the green material by increased titanium and iron, and also lower fluorine and increased boron. According to the original report, "The coexisting (equilibrium) pyroxenes also differ markedly from each other in manganese content and especially aluminum content. It should be noted that the association of brown idocrase + high-aluminum pyroxene is later than that of green idocrase + low-aluminum Mn-pyroxene + grossular. They replace each other according to the degree of lowering of the chemical potential of fluorine and increase in the potential of boron in the solutions." (Nekrasov 1971).

Figure 25. A classic skarn assemblage from Lake Jaco, Coahuila, Mexico: a sharp pink grossular crystal about 4 cm wide with yellowish-brown massive to crystalline vesuvianite in a matrix of altered limestone. *RJL3285*

Figure 26. Lustrous brown crystals lying flat in skarn from Italian Mountain, Taylor Park, Colorado. *RJL3379*

38 Formation and Geochemistry: Vesuvianites in Metamorphic Rocks

Veins in serpentinites represent another major source of fine vesuvianite crystals. In the area of Zermatt, Switzerland, Alpine fissures in metarodingites were formed in medium grade environments (300 – 500°C, 3 – 5 kbar); the rodingite is hosted by antigorite serpentinite. Elsewhere in the Swiss and Italian Alps, vesuvianite is found in various serpentinite rocks representing metamorphic grades from 200 to 800°C and 0 to 6.5 kbar. Many of these vein deposits are believed to have involved hydrothermal conditions. The premier North American vesuvianite locale, Jeffrey quarry, Asbestos, Quebec, Canada, represents hydrothermal veins (300 – 400°C, 0 – 2 kbar) in asbestos serpentinite. Other serpentinite occurrences include the Zhob Valley, Pakistan; the Bazhenovskoye deposit, near Asbest, Ural Mts., Russia; and the Mt. Belvidere asbestos quarry, Lowell, Vermont.

Figure 27. Green elongated prismatic crystal from Mt. Belvidere quarry, Lowell, Vermont. Note the similarity to some of the crystals from the Jeffrey quarry. *RJL3228*

Figure 28. Prismatic single crystal about 2 cm tall, in subtle shades of pink and green, from Jeffrey quarry, Asbestos, Quebec. *RJL3296*

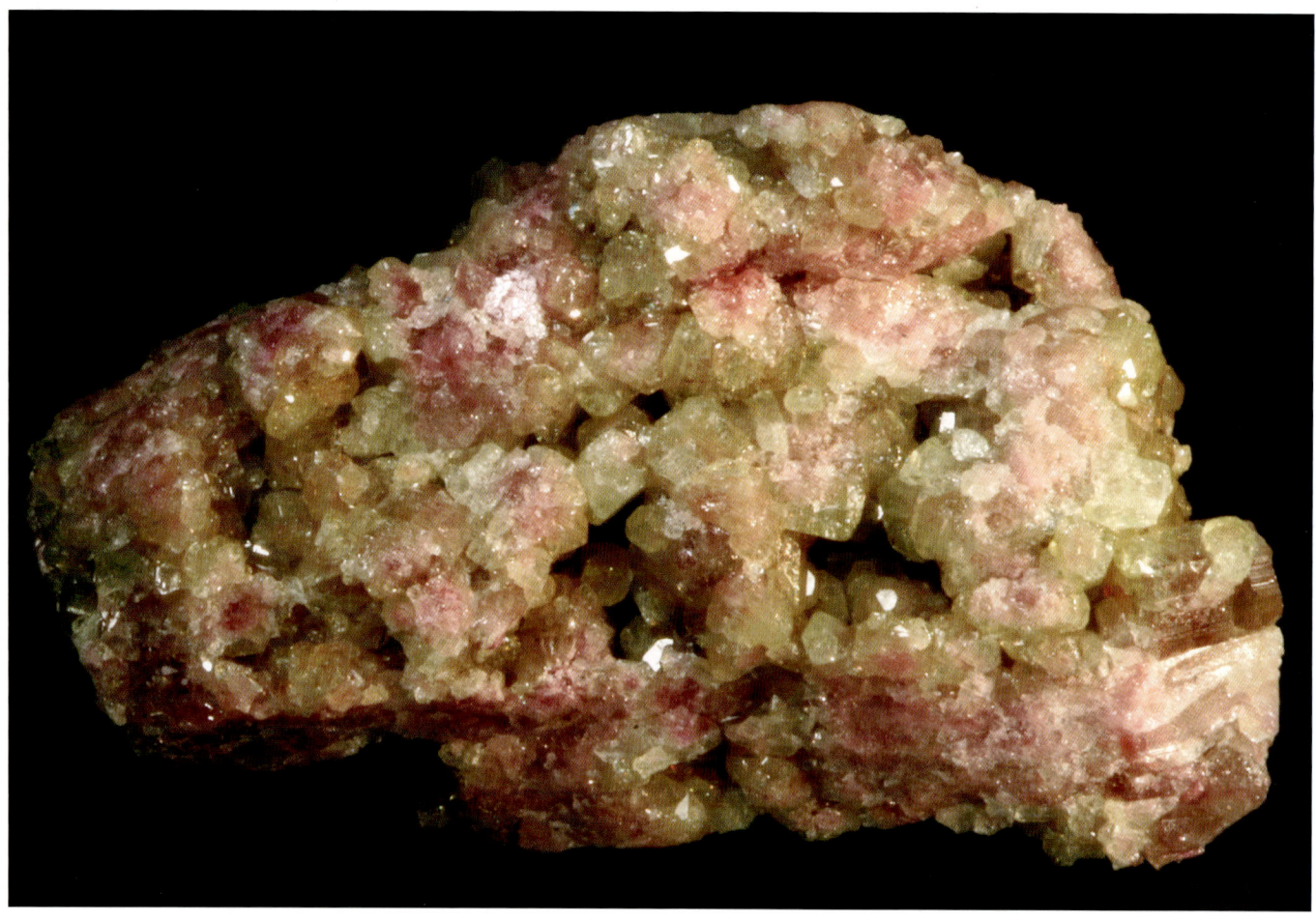

Figure 29. A mass of crystals 4 X 8 cm, in various colors ranging from pale green to deep pink, from Jeffrey quarry, Asbestos, Quebec. *RJL2607*

Hydrothermal deposits include those responsible for the recently described minerals fluorvesuvianite and manganvesuvianite. According to the original description (Britvin et al. 2003) fluorvesuvianite found at the Lupikko mine is a late-stage hydrothermal mineral whose origin is unrelated to that of the rock-forming vesuvianite that is widespread in the area. The paragenesis is hypothesized to have occurred in two stages: First, diopside-magnetite skarns were formed and these rocks contained vesuvianite as coarse-grained aggregates of cm-sized, brown prismatic crystals, typical of the vesuvianites often seen in skarn deposits. Later, these skarns were intensively altered by hydrothermal processes that formed abundant fluorite, calcite, and chlorite along with the fluorvesuvianite. At the Kalahari manganese fields, South Africa, manganvesuvianite was formed via hydrothermal alteration at 250 – 400°C of primary sedimentary and low-grade metamorphic manganese ores (Armbruster et al. 2002).

Figure 30. Elongated prismatic manganvesuvianite crystals forming a mass about 1 cm wide on rough octahedral hausmannite crystals, from South Africa. *RJL3241*

The Minerals

Vesuvianite was given its name by Werner in 1795 from the original locality on Mt. Vesuvius, Campania, Italy, where brown euhedral crystals are found in volcanic ejecta. There are well over 600 documented occurrences worldwide, and excellent crystals are known from many of them. Exhaustive locality listings can be found through various Internet resources and will not be repeated here. A few noteworthy localities are discussed below. With a little searching, the diligent collector should be able to find good examples of these specimens through commercial sources.

Figure 31. Typical brown vesuvianite crystals about 1 cm across, associated with small glassy white sodalite crystals in a vug in altered limestone from the type locale, Monte Somma, Vesuvius, Italy. *RJL2824*

The Alpine regions of Italy are rich in vesuvianite locales (Gramaccioli 1979), with a half-dozen in the Aosta Valley, over a dozen in the Piedmont, and four in Lombardy. Typically, rich green crystals are found in Alpine-type fissures in serpentinites. Similar deposits are known in nearby regions of Switzerland, particularly the area near Zermatt. Elsewhere in Europe, vesuvianite has been documented from skarns and contact metamorphic deposits at numerous locales in Germany, France, and Norway.

Figure 32. Simple striated prismatic crystal about 2 cm tall, without matrix, from an unspecified locality in Italy. *RJL3301*

Figure 33. Transparent green elongated prisms forming a plate about 2 cm across, from Bellecombe, Aosta Valley, Italy. *RJL1068*

Figure 34. Another example from Bellecombe, Aosta Valley, Italy: lustrous equant prismatic crystals to about 8 mm long. *RJL2713*

Figure 35. Another example from Bellecombe, Italy, displaying a more elongated prismatic habit. The crystal is about 3 cm tall and has an interesting area at the base where the crystal was deformed and later healed. *RJL3334*

Figure 36. Lustrous brown crystals to about 4 mm forming a thumbnail-sized plate, from Zermatt, Switzerland. This specimen is accompanied by an old handwritten label dated July 1837 – a testimony to the care with which this sample was curated for the last 170 years. *RJL2655*

Figure 37. Photomicrograph of an old specimen from Pusthertal, Tyrol, Austria, showing a colorful association of glassy yellow-green vesuvianite, red grossular var. *hessonite*, and green hexagonal chlorite crystals. *RJL3336*

Figure 38. Greenish, prismatic vesuvianite crystals about 7 mm long, from Hamrefjell, Norway. This sample shows the massive white calcite that frequently fills in between the vesuvianite crystals. For aesthetic reasons, the calcite is often removed with acid to better expose the crystals. *RJL3011*

50   The Minerals: Vesuvianite

Figure 39. Another vesuvianite from the same locale as the one shown in the previous figure. Note that no calcite is present; one can only wonder whether there was calcite among the crystals originally, and if so, whether it was removed by the natural leaching of water, or removed by acid treatment after it was collected. *RJL3378*

The Minerals: Vesuvianite   51

Figure 40. An interesting old-timer showing flattened prismatic crystals to about 1 cm long in a cavity in matrix. The locale was simply given as Germany, and the dealer noted that no additional information was available because the old label had disintegrated. This illustrates a valuable lesson for the conscientious collector: properly managing your collection includes *preserving the accompanying documentation* as well as caring for the physical wellbeing of the mineral specimen itself! *RJL3386*

Figure 41. Brownish vesuvianite forming radiating clusters of elongated prismatic crystals, from Hazlov u Chebu, near Eger, Czech Republic. This is the sort of material to which the varietal name *egeran* has historically been applied. *RJL3376*

Figure 42. Crude brown vesuvianite crystals with massive vesuvianite in skarn from Dognaczka, Hungary. This interesting older specimen likely dates from the 1930s. *RJL3387*

Sanford, Maine, has been known since the 1840s for very large, fine vesuvianites as well as excellent crystals of other species including grossular, titanite, meionite, actinolite, and scheelite. The locale was described by Kunz (1892) as follows: "About a mile and a half from Sanford, Me., idocrase occurs in unlimited quantities, one ledge, fully 30 feet wide, being made up entirely of massive idocrase, associated with quartz and occasionally with calcite, which fills the cavities containing the crystals. Some of the crystals are 7 inches long, and occasionally the smaller ones would afford fair gems." The mineralized zone here is a calc-silicate granofels completely surrounded by biotite granite, suggesting that the deposit is a roof pendant (Leavitt and Leavitt 1993). The locality has been referred to as the Goodall quarry, the Goodall farm mine, and the Webster quarry. The long productive history of the site, and the size and quality of its minerals, make Sanford, Maine, perhaps the classic American vesuvianite locale.

Figure 43. A large striated prismatic vesuvianite crystal in skarn from Sanford, Maine. *RJL2911*

Figure 44. An interesting association from Eden Mills, Vermont: a reddish-brown striated vesuvianite crystal about 5 X 20 mm, with small greenish yellow diopside crystals and orange grossular var. *hessonite*. RJL3391

The Bill Waley Allotment tungsten mine, Tulare County, California, is notable for an unusual mineral assemblage that includes some lustrous black vesuvianite whose dark color cannot be entirely explained on the basis of published analytical data. The deposit was formed at the contact between granodiorite and marble, where metasomatic replacement formed silicate and calc-silicate mineralizations. Some of the marble was later removed by groundwater, creating extensive brecciation along the contact zone. The mineral assemblage included loellingite, vesuvianite, calcite, epidote, wollastonite, pyrrhotite, and chalcopyrite. Both black and green vesuvianite has been found at the mine, and although the black crystals contain somewhat more total FeO + Fe_2O_3 (5.96 versus 4.25%) and TiO (0.39 versus 0.25%), these values are not unusually high. It was suggested that perhaps other impurities that were not specifically included in the analyses might be contributing to the unusually dark color (Crowley 1974).

Figure 45. A cluster of black vesuvianite from Bill Waley mine, California. Note that some references list this locale as Bill Waley Indian Allotment mine or Bill Waley Allotment mine. RJL3381

The Crestmore quarry, Riverside, California, is a geologically fascinating locality, where over two hundred mineral species have been identified in the skarn and ten minerals were described for the first time. It is perhaps best known in recent years for the large green clintonite crystals found in 2001, but rocks in the contact zone are also rich in vesuvianite; much of the vesuvianite is massive to granular, although prismatic crystals to 10 cm have been reported there (Forrester 2004).

Figure 46. A vesuvianite specimen collected ca. 1958 at Crestmore quarry, Riverside County, California. This sample shows the massive, granular habit of vesuvianite typically found at Crestmore. *RJL3281*

Figure 47. Green flattened vesuvianite crystals about 2 cm long, in altered limestone from Crestmore quarry, Riverside, California. The blue-gray material visible on the left is the remnants of calcite that was dissolved by acid cleaning in order to expose the vesuvianite crystals. *RJL3380*

The Jeffrey quarry, Asbestos, Quebec, (Grice and Williams 1979) has yielded fine vesuvianite crystal groups in a wide range of colors from rich green (chromian) to pale greenish-yellow (ferroan) to deep lavender (manganoan). However, even the most Mn-rich specimens from this locale lie within the composition range of vesuvianite (rather than manganvesuvianite *per se*). The quarry is presently closed to collecting but superb specimens were obtained there as recently as 2002 (Amabili, Miglioni, and Spertini 2004). "Thumbnail" collectors in particular will find a tremendous variety of sharp, colorful single crystals and groups widely available through mineral dealers, particularly those who have Canadian contacts. The photographs have been selected to show some of this variety.

Figure 48. A fine cabinet-sized example of green, Cr-rich vesuvianite from the Jeffrey quarry. The largest crystal is about 2 cm long. *RJL3385*

Figure 49. A vesuvianite crystal from the Jeffrey quarry showing an extreme example of color zoning, in which most of the crystal is pale green and there is a thin pink region at the termination indicating localized Mn enrichment. *RJL3348*

Figure 50. An interesting thumbnail-sized crystal group from Jeffrey quarry showing subtle color gradations from pale pink to pale green. *RJL3333*

Figure 51. Lustrous, equant crystals from Jeffrey quarry showing the typical range of colors from magenta to pink to green. *RJL3226*

The Minerals: Vesuvianite 61

Figure 52. A thumbnail-sized group of pale green crystals from Jeffrey quarry. *RJLjqd1*

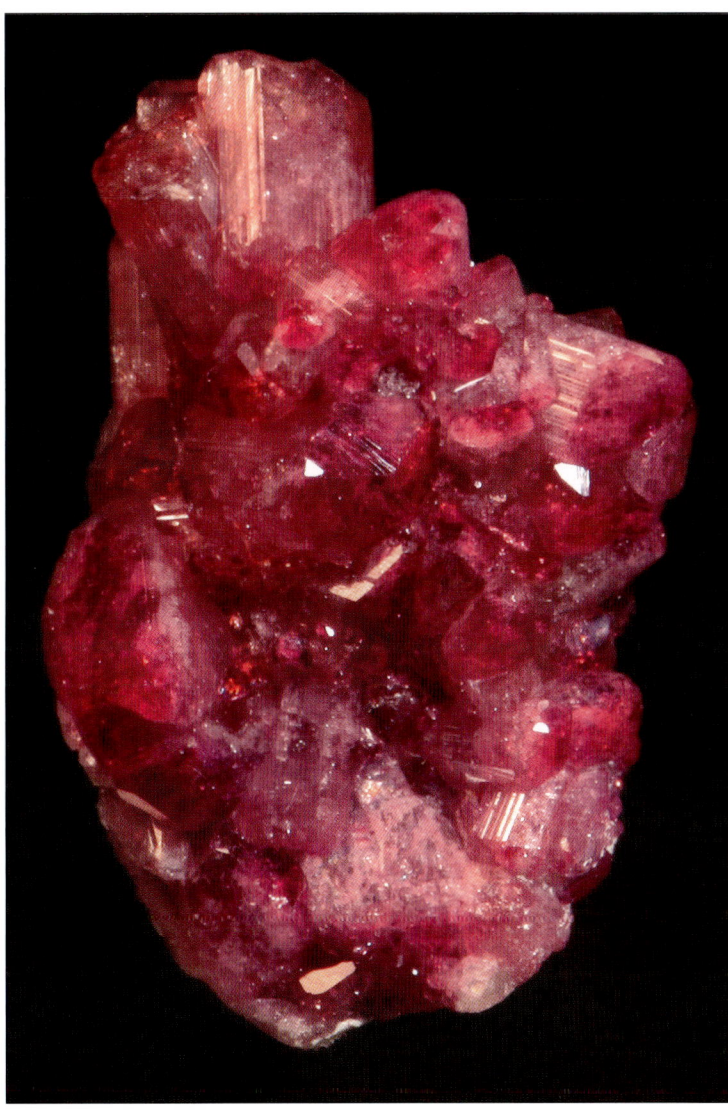

Figure 53. Another thumbnail from Jeffrey quarry consisting of bright magenta/pink crystals. *RJLjqc6*

Figure 54. An interesting doubly-terminated crystal that is mostly green with a very small pink, Mn-rich area at the lower end. *RJLjqa3*

Figure 55. Intense color zoning is exemplified by this crystal, which is pale yellow-green at each end and deep red in the center. By studying the color zoning, one can get a better understanding of how a particular crystal grew and how the solutions were changing with time during the growth process. *RJLjqb7*

Lake Jaco, Coahuila, Mexico, is a "classic" vesuvianite and garnet locale that has been a prolific source of tan, blocky vesuvianite crystals as well as pink to red grossular (for which it is presently more famous). Both Lake Jaco and the nearby village actually straddle the border between the states of Chihuahua and Coahuila, inspiring specimen labels that can be confusing at times. The actual mineral deposit lies to the east, in the Sierra de Cruces Range, Coahuila. The range is an alkaline diorite dome that intruded Cretaceous limestones. Contact metamorphism and metasomatism converted the limestones to skarns, creating a mineral assemblage including calcite, quartz, grossular, vesuvianite, wollastonite, and scapolite (Leuth and Jones 2003).

Figure 56. Dozens of intergrown brown vesuvianite crystals on matrix from Coahuila, Mexico. The blue-gray material surrounding the uppermost crystal is remnant calcite, most of which was removed by acid treatment. RJL2636

64 The Minerals: Vesuvianite

Figure 57. Cluster of cm-sized crystals on matrix from Coahuila, Mexico. *RJL3388*

Figure 58. A simple rectangular prism of vesuvianite with a dodecahedral grossular crystal attached to one corner, from Coahuila, Mexico. *RJL3232*

The Minerals: Vesuvianite 65

Figure 59. A very sharp prismatic crystal from Coahuila, Mexico. This example is interesting because although its morphology is not at all unusual for vesuvianites in general, it is somewhat less common at Lake Jaco than the simpler forms shown in the preceding photographs. *RJL3248*

Figure 60. Another single crystal from Lake Jaco, showing similar forms in slightly different proportions. *RJL3230*

Figure 61. A thumbnail-sized specimen from Lake Jaco, showing several lustrous intergrown vesuvianite crystals with tiny off-white grossular crystals. *RJL3231*

Veusivianite is found at several localities in the state of Rio Grande do Norte, Brazil, including at Malhada dos Angicos, Parelhas; and in the Barra Verde and Brejui tungsten mines.

Figure 62. Tabular brown vesuvianite crystals in matrix from Malhada dos Angicos, Parelhas, Rio Grande do Norte, Brazil. Specimen is about 2 cm tall. *RJL3335*

Figure 63. Equant brown vesuvianite crystal, about 1 cm tall, from Parelhas, Rio Grande do Norte, Brazil. *RJL3229*

Pakistan has become a major source of world-class gem crystals of many mineral species, and fine vesuvianites have been found in serpentinite in the Zhob Valley, Baluchistan. Brown euhedral crystals occur in Shalman, near the Khyber Pass on the Pakistan-Afghanistan border (this locale may be referred to as Warsak, which is the name of a large dam about six hours from Shalman). Fine root-beer-red crystals are also found in an alpine cleft deposit in Alchuri in the Shigar Valley (Blauwet 2006).

Gem-quality green to yellowish-brown vesuvianite is found in a skarn deposit in crystalline limestone near the Arusha to Nairobi Highway, between Kajiado and Namanga, Kenya (Keller 1992); faceted stones are occasionally available from gem dealers. Interestingly, a cut stone that likely came from this locality was studied by single-crystal neutron diffraction to elucidate the positions of hydrogen in the structure; the results ruled out the possibility of hydrogarnet-type substitution $[(H_4O_4)^{4-} \leftrightarrow (SiO_4)^{4-}]$, consistent with the results of IR spectroscopy on the same stone (Lager et al. 1999).

Figure 65. A thumbnail-sized cluster of dark brown crystals from Alchuri, Pakistan. Each crystal has a fibrous region at each end, creating a distinct chatoyancy. *RJL3340*

Figure 66. Dark brown equant crystals with white calcite on a schistose matrix, from Shalman (Warsak), Pakistan. Sample is about 5 cm across. *RJL2692*

Figure 64. A plate of small reddish brown crystals forming a solid plate about 5 cm across, from Alchuri, Pakistan. *RJL3339*

Particularly large crystals are found at Imilchil, and Boumia, Morocco; and at Fushan and elsewhere in Hebei Province, China.

As noted earlier, the vesuvianite structure can accommodate a wide variety of substitutions, and these impurities influence its color and other optical properties. It is not surprising, then, that many identifiable varieties have been named and described in the literature:

Figure 69. Brown vesuvianite from Sedi Bon Othman, Morocco. This interesting historical specimen represents the type of material described under the name *genevite* by Duparc and Gysin in 1927. The sample came from the School of Mines (Paris) and was originally in the collection of the BRGM (the French Bureau of Geological and Mineralogical Research) as the result of its work in colonial Africa. RJL3375

Figure 70. A large, flattened brown crystal of vesuvianite several cm across and about 1 cm thick, from Xintai, Hebei Province, China. RJL3010

Figure 67. A somewhat crude, but large, vesuvianite crystal about 7 cm tall, in matrix from Imilchil, Morocco. RJL2745

Figure 68. A lustrous brown crystal, about 3 × 6 cm, with other partial crystals and a bit of matrix, from Boumia, Morocco. RJL3390

Figure 71. Tabular vesuvianite, just under 3 cm tall, with small garnet crystals, from Han Dan mine, Hubei Province, China. RJL3130

Cerian vesuvianite has been found in altered greenstone at several outcrops near the Dallas benitoite mine, San Benito County, California, where it forms tiny striated prismatic crystals associated with black andradite var. *melanite*, perovskite, and clinochlore (Murdoch and Ingram 1966). The color ranges from amber to red to very dark red-brown. Crystals occasionally show a well-formed termination of the type expected in vesuvianites, but more often they show a poorly-formed or cavernous termination that appears more like a bundle or rim of needles around a concavity. The crystals are invariably quite small, but are well suited to preparing micromount specimens. Microprobe analysis indicated total rare-earth element (REE) content as high as 20 weight percent, with Ce dominant (Crook and Oswald 1979). This material also contains significant Ti (3 - 6% TiO_2) and rather low Al. Single-crystal X-ray diffraction studies on a small black crystal from the same locality provided insights into the complex coupled substitutions to accommodate the REE: 1) REE substitute for Ca; 2) Ti substitutes for Al on the general Al-Fe site; 3) the bridging O(11) site between the REE and Ti is occupied by oxygen, in contrast to other vesuvianites in which it is normally hydroxyl; and 4) to maintain overall charge neutrality Mg substitutes for Al in the Al-Fe site (Fitzgerald et al. 1987). Working collaboratively with Sharon Cisneros, Mineralogical Research Company, the author had an opportunity to study a large number of specimens from this locale; the variety of habits of the vesuvianite itself, and the interesting array of associated species, make these materials very desirable for the micromount enthusiast. Rare earth vesuvianite had previously been reported from hybrid rocks in the contact zone of Ayakhta granite intrusives in the Enisei Mountain Range, Russia (Orlov and Mart'yanov 1961). The Russian material had significantly lower REE content compared to the San Benito County samples (4% versus 17% RE_2O_3) and also contained small amounts of radioactive elements: 0.22% Th, 1.6×10^{-8} % Ra, and 4.8×10^{-2} % U. The crystals were similar to allanite in luster and color, and readily weathered to form red, orange, and white ochre-like surface coatings.

Figure 72. Photomicrograph of a small reddish brown Ce-rich vesuvianite crystal from San Benito County, California. Tiny black dodecahedral andradite crystals are associated. The vesuvianite crystal is about 3 mm long. *RJLsbc5*

Figure 73. Scanning electron micrograph of the specimen in the previous figure, viewed from a slightly different angle, showing one characteristic habit of this variety, in which the termination appears to be a bundle of needles. The dodecahedral shape of the garnet crystals can be seen clearly as well. *RJLsbc5*

Figure 74. Scanning electron micrograph of another Ce-rich vesuvianite from the same locale as the previous sample, showing the striations on the vesuvianite and, again, the intimate association with minute garnet crystals. *RJLsbc4*

Metamict vesuvianites have occasionally been found in alkaline igneous rocks. These materials contain uranium, thorium, and rare-earth elements (REE) and are of sufficient geological age that radioactive decay has disrupted their internal crystal structure. At Mt. Dakhunurskaya, southwest Tuva, Russia, a metamict vesuvianite containing about 1% U_3O_8, 0.5% ThO_2, and 0.67% REE_2O_3 occurs in an alkaline pegmatite with nepheline and zeolites. The mineral forms idiomorphic clusters of tabular crystals 3 to 7 cm long, which are black with greasy luster and conchoidal fracture; the overall appearance is described as similar to that of allanite (Kononova 1960). In the southeastern Seward Peninsula, Alaska, black tabular crystals to 1.5 cm are found in a radioactive syenite deposit, along with allanite, potassium feldspar, hornblende, and clinopyroxene. The crystals show some compositional zoning: in the metamict zones the crystals average about 0.8% U and about 2.7% Th, whereas in the non-metamict zones the respective concentrations are about half of these values (Himmelberg and Miller 1980). Black metamict vesuvianite has also been found in a skarn near Lake George, Colorado (Eby et al. 1993).

Figure 75. Photomicrograph of a specimen from San Benito County, California, showing another interesting association. Here, three elongated prismatic crystals of Ce-rich vesuvianite are associated with tiny black andradites and equant flesh-pink perovskite crystals. *RJL3244.*

Figure 76. Perovskite, titanian andradite, and Ce-rich vesuvianite on altered greenstone, from San Benito County, California. The red arrow at the top of the photo points to a perovskite crystal with a vesuvianite attached. *RJL3243*

An occurrence of *high-fluorine vesuvianite* "whiskers" was recently described from the Tas-Khayakhtakh Mountains of Polar Yakutia, Russia (Galuskin et al. 2003a). The reported fluorine content of this material, 3.7 F atoms per formula unit, seems to place it within the vesuvianite range, (compared to 4.6 F apfu in fluorvesuvianite), but it is fairly close to the boundary.

A *silicon-deficient vesuvianite* ("hydrovesuvianite") has recently been discovered in the Wilui region, Yakutia, Russia (Galuskin et al. 2003b) where it forms microscopic sheaf-like or spherulitic aggregates as a result of splitting of the {100} prism faces. The composition is explained on the basis of a hydrogarnet-type substitution of $(H_4O_4)^{4-}$ for $(SiO_4)^{4-}$. In this substitution, an SiO_4 unit is replaced by an H_4O_4 unit; the H_4O_4 unit is a tetrahedron but it is vacant at the center and the four H^+ ions (protons) are positioned slightly above the tetrahedral faces (Armbruster and Gnos 2000a). Based on the reported compositions, the material does not have enough of this substitution to warrant species status, and is therefore considered a variety of vesuvianite.

Figure 77. Photomicrograph of the specimen in the previous figure, showing a closer view of a striated red vesuvianite crystal on a perovskite crystal. *RJL3243*

Figure 78. Photomicrograph of a deep red, transparent prismatic crystal with a smooth, flat termination; among the rare-earth vesuvianites collected in San Benito County, this habit is less common than the striated crystals shown in the preceding figures. This sample was collected by Ed Oyler at a site about a mile south of the Gem mine, ca. 1958. *RJL3245*

Cyprine, a blue-green vesuvianite containing about 0.8 percent copper, is found in a hydrothermal-pegmatitic deposit at Øvstebo farm, Sauland, Telemark, Norway, associated with epidote, grossular, and massive pink zoisite (*thulite*). Beautiful specimens have also been found at the Franklin mine, Sussex County, New Jersey, where the cyprine is intimately intergrown with willemite and other species. Analysis of the Franklin material indicated 1.2% copper (Shannon 1922). Detailed crystal structure analysis was reported by Fitzgerald, Rheingold, and Leavens (1986).

Figure 79. Fine-grained pale blue vesuvianite var. cyprine with white calcite and willemite, and brown andradite, from Franklin, New Jersey. RJL3252

Figure 80. Small blue crystals of vesuvianite var. *cyprine* with colorless quartz and pink zoisite var. *thulite*, from Telemark, Norway. RJL3283

Chrome-vesuvianite, a bright green variety, is known from Xanthi, Greece, from the Yerington district, Nevada, and from several other localities; it is occasionally used for faceting rough.

Californite is a massive mixture of vesuvianite and grossular sometimes called California Jade, which can be made into beautiful cabochons. The best material has good color and translucency and is certainly as attractive as jade when used in fine jewelry.

Figure 81. A cm-sized green mass of crudely crystalline chrome-vesuvianite in matrix, from the Singatse Range, near Ludwig, Lyon County, Nevada.

Figure 82. Stream-worn hand specimen of collector-grade *californite* rough from the South Fork Mining claim; compared to the better gem grades, this type of material likely contains somewhat more grossular and somewhat less vesuvianite.

Beryllian vesuvianite has been noted from Franklin, New Jersey. The presence of beryllium was discovered around 1929 in an unidentified silicate material, which later optical and crystallographic analysis showed to be vesuvianite. This variety forms slender brown prismatic crystals embedded in a granular mixture of willemite, brown garnet, leucophoenicite, barite, and other species (Palache and Bauer 1930; Palache 1935).

Vesuvianite containing significant chlorine has been noted from rodingite-like metamorphic rocks in the Wilui River region, Russia, where it occurs in association with clinochlore, diopside, calcite, serpentine, and garnets of the hydrogrossular-andradite series. Chlorine tends to concentrate in the <001> and <101> growth sectors of tiny (50 μm) Si-deficient vesuvianite crystals and spherules. Based on that study, the authors raise a number of interesting arguments as to just how much Cl might be needed to define "chlorvesuvianite" as a valid species, and they rightly suggest that "chlorvesuvianite" needs further investigation (Galuskin, Galuskina, and Dzierzanowski 2005).

Figure 83. Beryllian vesuvianite forming a brown crystalline mass about 1 cm across in leucophoenicite, from Franklin, New Jersey. *RJL3377*

Figure 84. A closer view of the sample in the previous figure, showing small crude prismatic crystals of beryllian vesuvianite. *RJL3377*

Vesuvianite occasionally replaces other minerals, in some cases yielding recognizable pseudomorphs. In a zoned skarn in the Christmas Mountains, Texas, fibrous vesuvianite has partially or completely replaced euhedral prismatic grains of melilite, accurately preserving both the grain habit and the internal cleavages of the melilite (Joesten 1974). Similar alteration of earlier melilite skarns can be seen in rodingite-like rocks in the Wilui River region (Galuskin, Galuskina, and Dzierzanowski 2005). For pseudomorph collectors, perhaps the most interesting example is found at Franklin, New Jersey, where brown vesuvianite can be found replacing small crystals of mica (presumably phlogopite).

Figure 85. Small (1-2 mm) pods of brown vesuvianite forming pseudomorphs after mica crystals, from Franklin, New Jersey. *RJL3382*

Fluorvesuvianite is defined by the dominance of F at the **W** site in the general formula given above. The type material was discovered in 1999 in the abandoned Lupikko iron mine, Pitkäranta, Karelia, Russia. The mineral forms radiating aggregates of transparent acicular crystals in/on calcite in cavities in a chloritized diopside skarn, associated with sphalerite and clinochlore. The crystals range in size up to about 1.5 cm long and 5-20 μm thick. At the type locale, rock-forming vesuvianite is common in the mine. The fluorvesuvianite is a late-stage hydrothermal mineral whose formation appears to be unrelated to the rock-forming vesuvianite, but rather is attributed to late carbonate and fluorine metasomatism of the Lupikko ore bodies at temperatures below about 300°C (Britvin et al. 2003).

Figure 86. Photomicrograph of colorless, acicular fluorvesuvianite from the type locale at Lupikko iron mine, Pitkäranta, Karelia, Russia. *RJL2830*

Manganvesuvianite was recently described from the N'Chwaning II mine and the Wessels mine in the Kalahari manganese fields, South Africa (Armbruster et al., 2002; Armbruster and Gnos 2000a). The mineral was formed by hydrothermal alteration (250°–400°C) of primary sedimentary and low-grade metamorphic ores, which created mineralized zones along fault planes and lenses within the manganese ore beds or as veins and vug linings. It forms dark prismatic crystals up to about 1.5 cm long; larger crystals appear nearly black, but in strong transmitted light they are dark red-purple. In some areas of N'Chwaning II, smaller crystals of manganvesuvianite are densely intergrown with grossular or xonotlite, forming massive lenses within the manganese ore.

Figure 87. Photomicrograph of deep red elongated prismatic manganvesuvianite on black hausmannite from the type locale, N'Chwaning II mine, South Africa. *RJL3240*

Figure 88. Deep red prismatic manganvesuvianite crystals to about 1 cm long, completely covering matrix, from the type locale, N'Chwaning II mine, South Africa. *RJL2663*

Figure 89. An example of the "rock-forming" habit of manganvesuvianite in which the mineral forms a nearly solid mass of crystals intergrown with andradite and other species. *RJL3278*

Figure 90. A skeletal octahedron of hausmannite about 15 mm wide, with prismatic manganvesuvianite and pinkish manganoan calcite, a typical association found at N'Chwaning, South Africa. *RJL3242*

The Minerals: Wiluite

Wiluite was known for more than two hundred years from the Wilui River region, Sakha Republic, Russian Federation. It was long regarded as a boron-rich vesuvianite, but detailed crystallographic studies (Groat, Hawthorne, and Ercit 1994) determined that B occupies two distinctive sites in the structure. When these sites are more than half occupied, a separate species name is appropriate, and thus wiluite was formally described as a "new" species based on material from the original locality (Groat et al. 1998). Recent work has shown that high-B vesuvianites from Ariccia, Latium, Italy, are wiluite (Bellatreccia et al. 2005). Samples that appear to fall within the definition of wiluite have also been reported from Templeton Township, Quebec, Canada, and from the Bill Waley mine, Tulare County, California (Groat et al. 1996).

Figure 91. Three wiluite crystals from the type locale, Wilui River region, Sakha Republic, Russian Federation. [Note that a variety of names may be found for this locale on older labels, such as "Yakutia, Siberia," Yakutia being the old name of what is now the Sakha Republic. Other transliterations of the river's name, such as "Vilyuy," will also be seen in the literature.]

Figure 92. Brown prismatic wiluite crystals, each about 1 cm long, in matrix from the type locale. *RJL3280*

Figure 93. Intergrown crystals of wiluite from the type locale. The largest crystal is over 3 cm tall; note the surface figures on the prism faces and compare to the morphology drawings shown earlier. *RJL3279*

Figure 94. A dark purple-brown wiluite crystal, about 2 cm tall, in matrix from Ariccia, Italy. *RJL2934*

References

Amibili, M., A. Miglioli, and F. Spertini. 2004. Recent discoveries at the Jeffrey mine, Asbestos, Quebec. *Mineralogical Record* 35 (2): 123-35.

Arem, J. E. 1973. Idocrase (vesuvianite) – a 250-year puzzle. *Mineralogical Record* 4 (4): 164-74.

Arem, J. E. 1977. *Color Encyclopedia of Gemstones.* New York: Van Nostrand Reinhold, 149 pp.

Arem, J. E., and C. W. Burnham. 1969. Structural variations in idocrase. *American Mineralogist* 54: 1546-50.

Armbruster, T., and E. Gnos. 2000. P4/n and P4nc long range ordering in low-temperature vesuvianites. *American Mineralogist* 85: 563-69.

Armbruster, T., and E. Gnos. 2000a. Tetrahedral vacancies and cation ordering in low-temperature Mn-bearing vesuvianites: Indication of a hydrogarnet-like substitution. *American Mineralogist* 85: 570-77.

Armbruster, T., E. Gnos, R. Dixon, J. Gutzmer, C. Hejny, N. Dobelin, and O. Medenbach. 2002. Manganvesuvianite and tweddillite, two new Mn^{3+}-silicate minerals from Kalahari manganese fields, South Africa. *Mineralogical Magazine* 66 (1), 137-50.

Bauer, M. 1904. *Precious Stones* (1968 reprint). New York: Dover Publications, 627 pp.

Bellatreccia, F., F. Camara, L. Ottolini, G. Della Ventura, G. Cibin, and A. Mottana. 2005. Wiluite from Ariccia, Latium, Italy: occurrence and crystal structure. *Canadian Mineralogist* 43: 1457-68.

Blauwet, D. 2006. Famous mineral localities: Alchuri, Shigar Valley, Northern Areas, Pakistan. *Mineralogical Record* 37 (6): 513-40.

Britvin, S. N., A. A. Antonov, S. V. Krivovichev, T. Armbruster, P. C. Burns, and N. V. Chukanov. 2003. Fluorvesuvianite, $Ca_{19}(Al,Mg,Fe^{2+})_{13}[SiO_4]_{10}[Si_2O_7]_4O(F,OH)_9$, a new species from Pitkäranta, Karelia, Russia: description and crystal structure. *Canadian Mineralogist* 41:1371-80.

Chamberlain, S. C. 1978. The photographic record: The Waddell collection. *Mineralogical Record* 9 (2): 91-4.

Crook, W. W., and S. G. Oswald. 1979. New data on cerian vesuvianite from San Benito County, California. *American Mineralogist* 64: 367-68.

Crowley, J. A. 1974. Loellingite and black vesuvianite from Tulare County, California. *California Geology* 27: 224-26.

Deer, W. A., R. A. Howie, and J. Zussman. 1982. *Rock-Forming Minerals*, Vol. 1A Orthosilicates. London: Longman's.

Eby, R. K., J. Janeczek, R. C. Ewing, T. S. Ercit, L. A. Groat, B. C. Chakoumakos, F. C. Hawthorne, and G. R. Rossman. 1993. Metamict and chemically altered vesuvianite. *Canadian Mineralogist* 31: 357-69.

Fitzgerald, S., P. B. Leavens, and J. A. Nelen. 1992. Chemical variation in vesuvianite. *Mineralogy and Petrology* 46: 163-78.

Fitzgerald, S., A. L. Rheingold, and P. B. Leavens. 1986. Crystal structure of a Cu-bearing vesuvianite. *American Mineralogist* 71: 1011-14.

Fitzgerald, S., A. L. Rheingold, and P. B. Leavens. 1986a. Crystal structure of a non-P4nnc vesuvianite from Asbestos, Quebec. *American Mineralogist* 71: 1483-88.

Fitzgerald, S., P. B. Leavens, A. L. Rheingold, and J. A. Nelen. 1987. Crystal structure of a REE-bearing vesuvianite from San Benito County, California. *American Mineralogist* 72: 625-28.

Forrester, C. 2004. Large clintonite crystals from the Crestmore Quarry, Riverside, California. *Mineralogical Record* 35 (4): 325-330.

Galuskin, E. V., T. Armbruster, A. Malsy, I. O. Galuskina, and M. Sitarz. 2003a. Morphology, composition, and structure of low-temperature P4/nnc high-fluorine vesuvianite whiskers from Polar Yakutia, Russia. *Canadian Mineralogist* 41: 843-56.

Galuskin, E. V., I. O. Galuskina, G. Bzowska, and M. Outrequin. 2003b. Autodeformation mechanism of splitting of Si-deficient vesuvianite crystals. *Mineralogical Society of Poland – Special Papers* 22: 51-3.

Galuskin, E. V., I. O. Galuskina, and P. Dzierzanowski. 2005. Chlorine in vesuvianites. *Mineralogia Polonica* 36 (1): 51-61.

Gibson, R. L., T. Wallmach, and D. de Bruin. 1995. Complex zoning in vesuvianite from the Canigou Massif, Pyrenees, France. *Canadian Mineralogist* 33: 77-84.

Gnos, E., and T. Armbruster. 2006. Relationship among metamorphic grade, vesuvianite "rod polytypism", and vesuvianite composition. *American Mineralogist* 91: 862-70.

Gnos, E., and T. Armbruster. 2007. Vesuvian. *Schweitzer Strahler* 2007 (1): 13-21.

Goldschmidt, V. 1916. *Atlas der Krystallformen* [see Facsimile Reprint in Nine Volumes (1986) by the Rochester Mineralogical Symposium].

Gramaccioli, C. M. 1979. Minerals of the Alpine rodingites of Italy. *Mineralogical Record* 10 (2): 85-9.

Grice, J. D., and R. Williams. 1979. Famous mineral localities: the Jeffrey mine, Asbestos, Quebec. *Mineralogical Record* 10 (2): 69-80.

Groat, L.A., F. C. Hawthorne, and T. S. Ercit. 1992a. The chemistry of vesuvianite. *Canadian Mineralogist*, 30: 19-48.

Groat, L.A., F. C. Hawthorne, and T. S. Ercit. 1992b. The role of fluorine in vesuvianite; a crystal structure study. *Canadian Mineralogist*, 30: 1065-75.

Groat, L.A., F. C. Hawthorne, and T. S. Ercit. 1994. The incorporation of boron into the vesuvianite structure. *Canadian Mineralogist* 32: 505-23.

Groat, L.A., F. C. Hawthorne, and T. S. Ercit. 1994a. Excess Y-group cations in the crystal structure of vesuvianite. *Canadian Mineralogist* 32: 497-504.

Groat, L. A., F. C. Hawthorne, G. A. Lager, A. J. Schultz, and T. S. Ercit. 1996. X-ray and neutron crystal-structure refinements of a boron-bearing vesuvianite. *Canadian Mineralogist* 34: 1059-70.

Groat, L.A., F. C. Hawthorne, T. S. Ercit and J. D. Grice. 1998. Wiluite, $Ca_{19}(Al,Mg,Fe,Ti)_{13}(B,Al,\square)_5Si_{18}O_{68}(O,OH)_{10}$, a new mineral species isostructural with vesuvianite, from the Sakha Republic, Russian Federation. *Canadian Mineralogist* 36: 1301-4.

Groat, L.A., F. C. Hawthorne, T. S. Ercit, and A. Putnis. 1993. The symmetry of vesuvianite. *Canadian Mineralogist*, 31: 617-35.

Himmelberg, G. R., and T. P. Miller. 1980. Uranium- and thorium-rich vesuvianite from the Seward Peninsula, Alaska. *American Mineralogist* 65: 1020-25.

Joesten, R. 1974. Pseudomorphic replacement of melilite by idocrase in a zoned calc-silicate skarn, Christmas Mountains, Big Bend region, Texas. *American Mineralogist* 59: 694-99.

Keller, P. C. 1992. *Gemstones of East Africa*, 144 pp., Phoenix: Geoscience Press.

Kononova, V. A. 1960. On a metamict variety of vesuvianite from an alkaline pegmatite in southwest Tuva. *Dokl. Acad. Nauk SSSR, Earth-Sci. Sect.* 130: 129-32.

Kraft, J. L. 1947. *Adventure in Jade*. New York: Henry Holt and Co.

Kunz, G. F. 1892. *Gems and Precious Stones of North America*, Second Edition, (1968 reprint). New York: Dover Publications, 367 pp.

Lager, G. A., Q. Xie, F. K. Ross, G. R. Rossman, T. Armbruster, F. J. Rotella, and A. J. Schultz. 1999. Hydrogen-atom positions in P4/nnc vesuvianite. *Canadian Mineralogist* 37: 763-8.

Lauf, R. J. 2009. Collectors guide to the vesuvianite group. *Rocks & Minerals*, in press.

Leavitt, D. L., and N. J. Leavitt. 1993. Mineralogy of the Sanford vesuvianite deposit. *Mineralogical Record* 24 (5): 359-64.

Leuth, V. W., and R. Jones. 2003. Red grossular from the Sierra de Cruces, Coahuila, Mexico. *Mineralogical Record* 34 (6): 73-95.

Mandarino, J. A., and V. Anderson. 1989. *Monteregian treasures: the minerals of Mont Saint-Hilaire, Quebec.* 281 pp., Cambridge: Cambridge University Press

Manning, P. G. 1975. Charge-transfer processes and the origin of colour and pleochroism of some titanium-rich vesuvianites. *Canadian Mineralogist* 13: 110-16.

Manning, P. G. 1977. Charge-transfer interactions and the origin of color in brown vesuvianite. *Canadian Mineralogist* 15: 508-11.

Manning, P. G., and M. J. Trickler. 1975. Optical-absorption and Mössbauer spectral studies of iron and titanium site-populations in vesuvianites. *Canadian Mineralogist* 13: 259-65.

Murdoch, J., and B. L. Ingram. 1966. A cerian vesuvianite from California. *American Mineralogist* 51: 381-87.

Nekrasov, I. Ya. 1971. Features of tin mineralization in carbonate deposits, as in Eastern Siberia. *Internat'l Geol. Rev.* v. 13, no. 10, pp. 1532-42. [Trans. from *Sovietskaya Geologiya* 1970, no. 12, pp. 41-54.]

Ohkawa, M., A. Yoshiasa, and S. Takeno. 1992. Crystal chemistry of vesuvianite: Site preferences of square-pyramidal coordinated sites. *American Mineralogist* 77: 945-53.

Orlov, Yu. L., and N. N. Mart'yanov. 1961. Vesuvianite containing rare-earth elements, from the Enisei Mountain Range. *Trudy Mineralog. Muzeya Akad. Nauk SSSR* 11: 187 [see abstract 1961, *Chem. Abstr.* 55: 20803].

Palache, C. 1935. *The Minerals of Franklin and Sterling Hill Sussex County, New Jersey.* Geological Survey Professional Paper 180, U.S. Government Printing Office, 135 pp.

Palache, C. and L. H. Bauer. 1930. On the occurrence of beryllium in the zinc deposits of Franklin, New Jersey. *American Mineralogist* 15: 30-33.

Patel, S. C. 2007. Vesuvianite-wollastonite-grossular-bearing calc-silicate rock near Tatapani, Surguja district, Chhattisgarh. *J. Earth Syst. Sci.* 116 (2): 143-47.

Pavese, A., M. Prencipe, M. Tribaudino, and S. S. Aagaard. 1998. X-ray and neutron single-crystal study of P4/n vesuvianite. *Canadian Mineralogist* 36: 1029-37.

Robinson, G. and S. C. Chamberlain. 1982. An introduction to the mineralogy of Ontario's Grenville province. *Mineralogical Record* 13 (2): 71-86.

Rucklidge, J. C., V. Kocman, S. H. Whitlow, and E. J. Gabe. 1975. The crystal structures of three Canadian vesuvianites. *Canadian Mineralogist* 13: 15-21.

Shannon, E. V. 1922. Note on the cyprine from Franklin Furnace, New Jersey. *American Mineralogist* 7: 140-42.

Tanaka, T., M. Akizuki, and Y. Kudo. 2002. Optical properties and crystal structure of triclinic growth sectors in vesuvianite. *Mineralogical Magazine* 66 (2): 261-74.

Truebe, H. A. 1984. Minerals of the Italian Mountain area, Colorado. *Mineralogical Record* 15 (2): 75-88.

Veblen, D. R. and M. J. Wiechmann. 1991. Domain structure of low-symmetry vesuvianite from Crestmore, California. *American Mineralogist* 76: 397-404.

Yoshiasa, A. and Matsumoto, T. 1986. The crystal structure of vesuvianite from Nakatatsu mine: reinvestigation of the cation site-populations and of the hydroxyl groups. *Mineralogical Journal* 13 (1): 1-12.

More Schiffer Earth Science Monograph Titles

www.schifferbooks.com

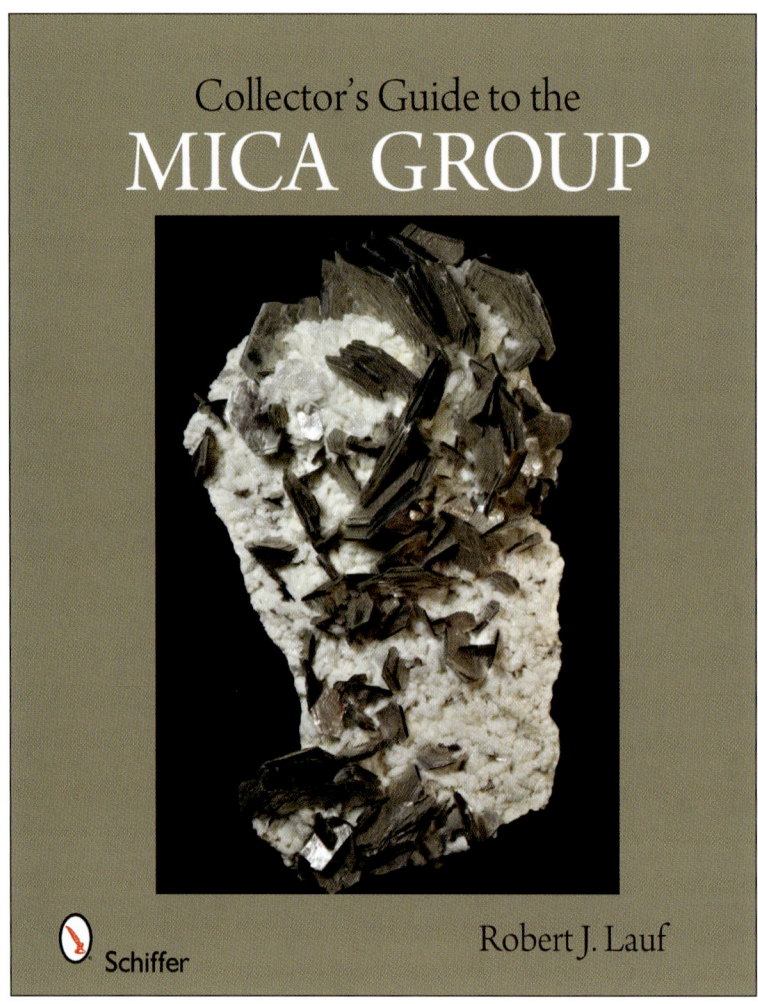

Collector's Guide to the Mica Group. Robert J. Lauf. Mica is a broad term encompassing about forty minerals, ranging from the common to the rare, many at times forming excellent crystals jewelers use. This book feaures examples recently described among the 115 striking color photos and electron micrographs that illustrate the text. A detailed entry for each type includes information on where each is found, associations of micas with other minerals, pseudomorphs (minerals that masquerade as mica), and micas that fluoresce under UV light. This fascinating guide is for those interested in minerals.

Size: 8 1/2" x 11"   115 color photos   96pp.
ISBN: 978-0-7643-3047-6   soft cover $19.99

Collector's Guide to the Epidote Group. Robert J. Lauf. Over 90 striking color photos display minerals of the epidote group, well known to mineral collectors for their rich colors and the many interesting minerals with which they occur. Lapidary artists also value epidote, particularly in the form of unakite, and precious or semiprecious varieties of the related mineral zoisite, including thulite and tanzanite, some which have inclusions that allow them to be cut into popular catseyes. This informative book provides all presently known species, detailed entries for each of the eighteen minerals, and extensive locality information. This book will be of interest to those interested in developing a better understanding of silicate minerals.

Size: 8 1/2" x 11"   92 color photos   96pp.
ISBN: 978-0-7643-3048-3   soft cover $19.99

Collector's Guide to Fluorite. Arvid Eric Pasto. Fluorite is found everywhere, has been important to industry for centuries, and is a minor ornamental material as well. Fluorite presents a fascinating array of colors, habits, and associated minerals and is widely available. See spectacularly large "museum quality" specimens that can be found. Fluorite presents a wealth of scientific opportunities to see crystallography, geochemistry, and solid-state physics at work in the natural world. It provides over 140 full-color examples and extensive references to the formation and geographic locations of fluorite. This book is essential for everyone with a passion for minerals.

Size: 8 1/2" x 11"   143 color   96pp.
                       photos, 10 illus.
ISBN: 978-0-7643-3193-0   soft cover $19.99

Other Schiffer Titles

www.schifferbooks.com

Introduction to Radioactive Minerals. Robert J. Lauf. Collectors have long admired uranium and thorium minerals for their brilliant colors, intense ultra-violet fluorescence, and rich variety of habits and associates. Radioactive minerals are also critically important as our source of nuclear energy. Understanding them is crucial to the safe disposal of radioactive waste. This book provides a systematic overview of the mineralogy of uranium and thorium, generously illustrated with nearly 200 color photos and electron micrographs of representative specimens. Includes an historical discussion of the discovery of radioactive elements and the development of uranium and thorium ore deposits, a discussion of the geochemical conditions that produce significant deposits, and a description of important localities, their geological setting and history. Major occurrences of interest to mineral collectors are arranged geographically. The minerals are arranged systematically, to emphasize how they fit into chemical groups, and for each group a few minerals are selected to illustrate their formation and general characteristics. With the resurgence of interest in nuclear power, this book is an invaluable guide for mineral collectors as well as nuclear scientists and engineers interested in radioactive deposits.

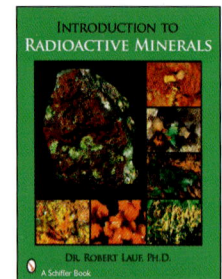

Size: 8 1/2" x 11" 196 color & b/w photos 144pp.
ISBN: 978-0-7643-2912-8 soft cover $29.95

Collecting Fluorescent Minerals. Stuart Schneider. Seeing fluorescent minerals up close for the first time is an exciting experience. The colors are so pure and the glow is so seemingly unnatural, that it is hard to believe they are natural rocks. Hundreds of glowing minerals are shown, including Aragonite, Celestine, Feldspar, Microcline, Picropharmacolite, Quartz, Spinel, Smithsonite, plus many more. But don't let the hard-to-pronounce names keep you away. Over 800 beautiful color photographs illustrate how fluorescent minerals look under the UV light and in daylight, making this an invaluable field guide. Included are values, a comprehensive resources section, plus helpful advice on caring for, collecting, and displaying minerals. The field of collecting fluorescent minerals is relatively new and this is one of the most complete references available.

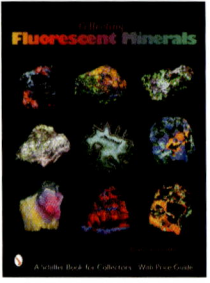

Size: 8 1/2" x 11" 846 color photos 192pp.
ISBN: 0-7643-2091-2 soft cover $29.95

The World of Fluorescent Minerals. Stuart Schneider. The rich and diverse world of fluorescent minerals is explored in this sweeping survey. Breathtakingly pure colors, with their ethereal glow, immediately capture your attention. Did you know that color television is a result of the study of fluorescing minerals? Fresh finds of fluorescent minerals are showing up regularly around the globe, and their collection is an entertaining and popular pasttime. To help the collector, over 825 photos display the minerals both as they might be found in daylight and in under the effects of ultraviolet light. Written for the collector and the merely curious, this pictorial reference will enrich your collecting experience with its informative text. It is an essential source for enjoying and identifying fluorescent minerals.

Size: 8 1/2" x 11" 825 color photos 176pp.
ISBN: 0-7643-2544-2 soft cover $29.95

Schiffer books may be ordered from your local bookstore, or they may be ordered directly from the publisher by writing to:

Schiffer Publishing, Ltd.
4880 Lower Valley Rd.
Atglen, PA 19310
(610) 593-1777; Fax (610) 593-2002
E-mail: Info@schifferbooks.com

Please visit our web site catalog at *www.schifferbooks.com* or write for a free catalog. Please include $5.00 for shipping and handling for the first two books and $2.00 for each additional book. Full-price orders over $150 are shipped free in the U.S.

Printed in China